SpringerBriefs in Applied Sciences and Technology

SpringerBriefs present concise summaries of cutting-edge research and practical applications across a wide spectrum of fields. Featuring compact volumes of 50 to 125 pages, the series covers a range of content from professional to academic.

Typical publications can be:

- A timely report of state-of-the art methods
- An introduction to or a manual for the application of mathematical or computer techniques
- A bridge between new research results, as published in journal articles
- A snapshot of a hot or emerging topic
- An in-depth case study
- A presentation of core concepts that students must understand in order to make independent contributions

SpringerBriefs are characterized by fast, global electronic dissemination, standard publishing contracts, standardized manuscript preparation and formatting guidelines, and expedited production schedules.

On the one hand, **SpringerBriefs in Applied Sciences and Technology** are devoted to the publication of fundamentals and applications within the different classical engineering disciplines as well as in interdisciplinary fields that recently emerged between these areas. On the other hand, as the boundary separating fundamental research and applied technology is more and more dissolving, this series is particularly open to trans-disciplinary topics between fundamental science and engineering.

Indexed by EI-Compendex, SCOPUS and Springerlink.

More information about this series at http://www.springer.com/series/8884

Antonella Ingenito

Subsonic Combustion Ramjet Design

 Springer

Antonella Ingenito
School of Aerospace Engineering
Sapienza University of Rome
Rome, Italy

ISSN 2191-530X   ISSN 2191-5318 (electronic)
SpringerBriefs in Applied Sciences and Technology
ISBN 978-3-030-66880-8   ISBN 978-3-030-66881-5 (eBook)
https://doi.org/10.1007/978-3-030-66881-5

This Springer imprint is published by the registered company Springer Nature Switzerland AG
The registered company address is: Gewerbestrasse 11, 6330 Cham, Switzerland

Contents

1   **Introduction** 1
     References 3

2   **Fundamentals of Ramjet Engines** 5
     References 7

3   **Design of Supersonic/Hypersonic Vehicles** 9
     3.1   Küchemann tau Figure of Merit 9
     3.2   Breguet Range for Airbreathing Supersonic/Hypersonic
         Vehicles 10
     3.3   Hypersonic Convergence Equations 12
         3.3.1   Weights and Volumes Hypersonic Convergence 12
     References 17

4   **Ramjet Engines Performance** 19
     4.1   Ideal Ramjet Cycle 19
     4.2   Real Ramjet Cycle 27
     4.3   Thrust Parameters and Efficiencies 31
     References 34

5   **Ramjets Fuels** 35
     5.1   Proprieties of Ramjet Fuels 35
     5.2   Fuel Performance Comparison 37
     5.3   Ignition Delay Time for Ramjet Fuels 39
     5.4   Flame Speed for Ramjet Fuels 42
     References 45

6   **Flameholder Design Guidelines** 47
     6.1   Flameholder Geometries 47
     6.2   Theoretical Analysis of V-type Flameholder Geometry ... 50
     6.3   Flame Stability 54
     6.4   Investigation of Flameholder Blockage Area Ratio 56
     6.5   Some Examples: Multistage V-gutter Flameholders 56
         6.5.1   Two Parallel Lines of Gutters 58

    6.5.2   Three Parallel Lines of Gutters 60
    6.5.3   Staggered Gutters 60
    6.5.4   Three Rows of Staggered Gutters 61
    6.5.5   Other Configurations 62
    6.5.6   Rake- and Gutter-Type Flameholders 62
    6.5.7   Grid Type Flameholder 64
    6.5.8   Annular Flameholders 66
    6.5.9   Investigation of Flameholder V Angle and Width 66
  References 73

7 **Injector Design Guidelines** 75
  7.1   Investigation of Injector Number, Size and Radial
        Distribution in Flame Holders 75
  7.2   Droplets Evaporation Time 77
  7.3   Fuel Jet Penetration and Breakup 78
  7.4   Calculation of Fuel Spatial Distribution with the Distance
        from the Flameholder 81
  7.5   Experimental Analysis of Injector Radial Distance
        from the Outer Wall 82
  References 85

8 **Combustor Design Guidelines** 87
  8.1   Cylindrical Combustion Chamber Sizing 87
  Reference 89

9 **Igniter Design Guidelines** 91
  9.1   Flammability Limits 91
  9.2   Ignition Energy 91
  References 93

10 **Step by Step RJ Design Methodology** 95

11 **Material Selection for Ramjet Engines** 99
  11.1   Operation Temperatures for Materials 99
  11.2   Ramjet Material Requirements 105
  References 113

12 **Conclusions** 115

Nomenclature

A	Arrhenius constant
A	Area [m^2]
Af	flame holder open area [m^2]
Afh	flame holder area [m^2]
alpha_fh	30°, 45°, 60° V gutter angle
b_fh	flameholder width [m]
Bq	dimensionless number transfer [-]
$B_L$mix	mixing efficiency [-]
BR	blockage ratio (1-Af) [-]
c	static pressure drop coefficient defined as $(p_1\text{-}p_2)/q_1$ [-]
C	Total-pressure-drop
Cp	specific heat [J/kg/K]
d	diameter [m]
h	penetration of liquid jet in Xdirection
H	height from bottom combustor wall
Hn	distance from bottom combustor wall to exit of fuel injector
L	distance from center of fuel injector to wind-ward edge of gutter in Zdirection
Ma	Mach number of air stream
[O_2]	oxygen mole concentration
Ps	static pressure
q	fuel-to-air momentum ratio
Ta	static air temperature
Tc	static temperature of combustion gas at combustor exit
Tf	fuel injection temperature
Uj	fuel injection velocity
Va	air velocity in Zdirection
X	distance from fuel injector exit in direction of liquid jet injection
Y	distance from center axis of combustor in direction perpendicular to X and Z axis

Z	distance from center of fuel injector exit in direction of airflow
Zg	distance from leeward edge of gutter in Z direction
Zh	distance from upstream side edge of fuel injector exit in direction of airflow (= Z+d/2)

Greek Symbol

η	combustion efficiency
ϱ	density
σ	total pressure loss at combustor exit normalized by total pressure at inlet
φ	equivalence ratio
D	diameter [m]
E	activation Energy [kcal/mol]
eta	combustion efficiency
F/A	fuel air ratio [-]
F/Ast	stoichiometic fuel air ratio [-]
hf	formation enthalpy of the fuel gas phase [J/mol]
Ia	Specific Thrust [s]
Isp	Specific impulse [s]
k	thermal diffusivity [W/m/K]
K	flow area contraction coefficient, experimentally found to vary from 0.8 to 0.95, [-]
Kev	evaporation rate constant [-]
Lacc	length required to ignite [m]
Lb_fh	flameholder span [m]
Lbreak	xb distance required for the jet to break from the injector [m]
Lcomb	assigned combustor length [m]
Lcombreq	length required for the combustion to complete (calculated) [m]
Levap	length required to vaporize [m]
L_fh_ign	xdistance between the injectors and the flameholder [m]
Lflame	length required to mix and complete combustion [m]
Lphi	distance from injectors at which stoichiometric fuel concentration are achieved [m]
m_0	air mass flow [kg/s]
m_f	fuel mass flow [kg/s]
M	Mach number [-]
MIE	Minimum ignition energy [J]
n	empirical constants for ignition delay calculation
ninj	injectors number [-]
P	pressure [atm]
q	fuel-to-air momentum ratio [-]
R	radius [m]
R	specific gas constant [J/kg/K]

Sl	laminar flame speed [m/s]
SMD	Souter Mean Diameter
St	turbulent flame speed [m/s]
St	Specific thrust [s]
t_{St}	time required for the flame to pass through the entire air-fuel mixture:
T	temperature [K]
T	Thrust [N]
Tb	parameter of Jet breakup
TSFC	Thrust specific fuel consumption [kg/h/N]
U	velocity [m/s]
utsql	turbulence intensity
We	Weber dimensionless number [-]
x	position where the local fuel air ratio is stoichiometric
xb	distance required for the jet to break from the injector [m]
z	flight altitude [m]

Greek Symbol

Φ	equivalence ratio [-]
Φl	lean blowout equivalence ratio [-]
@	heat capacity ratio [-]
@	injection angle
μ	viscosity [kg/m/s]
η	efficiency
ϱ	density[kg/m^3]
σ	surface tension [N/m]
τ	ignition delay time [s]
τ_{ab}	jet break time via aerodynamic forces
τ_{fb}	jet break time via fuel jet forces

Subscript and Superscript

a	air
'	ideal conditions
cc	combustion chamber
comb	combustion
ev	evaporation
f	fuel
fh	flameholder
in	inlet

inj	injector
m	main fuel
j	fuel jet
jet	fuel jet
LBO	lean blow out limit
r	real
res	residence
s	static
sto	storage
t	total
0	free stream conditions
1	ram compressor inlet
2,3	combustor inlet
4,7	combustor outlet
9	9 nozzle outlet

Chapter 1
Introduction

Abstract A Ramjet (RJ) is the simplest aeropropulsion system, consisting of convergent–divergent air inlet, to decelerate air entering the combustor, a combustor (including fuel injectors and flame anchoring devices), and a nozzle to accelerate the combustion products. A very extensive review of the ramjet history has been done by Fry and Waltrup [1, 2].

Development of viable ramjet took many years. In fact, ground tests of a full-scale Mach 2.35 ramjet-powered aircraft begun by 1938 [3, 4]. In the late '50s, numerous projects investigated the feasibility of implementing ramjet engines. In particular, Lockheed started systematic tests of ramjets under the experimental X-7 program, achieving increasing speed up to Mach 4.33. These efforts resulted in weapon systems such as the Boeing *Bomarc (U.S. Air Force)*, *Talos (U.S. Navy)*, and *Bloodhound (Great Britain)* antiaircraft missiles. The first Talos RIM-8 (ground-to-air), in service since 1952, incorporated a rocket booster to accelerate the missile up to the ramjet operating speeds. The Soviets also implemented ramjets in the 2K11 "Krug" (SA-4) and 2K12 "Kub" (SA-6) missiles. In Europe, *France* developed several operational ramjet missiles (*VEGA, CT-41, and SE 4400*) in the late 1950s and early 1960s.

With the rapid improving of the rocket technology, interest in ramjet engines decreased. Only Soviets continued to invest in hypersonic and toward the end of the 1970s, they built missiles like the P-270 Moskit and P-800 Onyks.

From 1980s to date, the ramjet engine received again attention. In Europe, France developed the operational Air-Sol Moyenne Portee-Ameliore (ASMP) and flight-tested the Missile Probatoire Stato Rustique (MPSR)/Rustiqueè. ASMP was 5.38 m long with a diameter of 300 mm and weighted 860 kg. It was a supersonic stand-off missile powered by a liquid-fuel (kerosene) ramjet. It flew at a speed of up to Mach 3, with a range between 80 and 500 km depending on flight profile. In 90s, France continued its long history of development activity in ramjets with activity on MARS, MPSR/Rustique, Anti-Navire Futur/Anti-Navire Nouvelle Generation ANF/ANNG, Vesta, and the ASMP-A. ASMP-A entered service in October 2009 with the Mirages and on July 2010 with the Rafales.

A. Ingenito, *Subsonic Combustion Ramjet Design*,
SpringerBriefs in Applied Sciences and Technology,
https://doi.org/10.1007/978-3-030-66881-5_1

The first-generation missile ramjets like RIM-8, 2K12, 2K11 and Sea Dart were all based on liquid ramjets and were very large. India and Russia also developed the BrahMos missile, operative since 2013. This missile is an updated version of the P-800 Onyks cruise missile; this is capable of speed close to Mach 3.0 and a range of about 300 km, with a weight of about 2.5 tons for the version launched from aircraft. For this reason, only few aircrafts like the Sukhoi Su-30MKI were able to carry it.

The introduction of solid propulsion [5–7] has allowed designers in Europe, Russia and China to extend the range of the missile from 100 to 200 km and to reach a maximum speed of Mach 4, without significantly increasing the diameter and length of the missile.

In Europe, the British solid-fueled Sea Dart Meteor missile (MBDA UK) was capable of reaching $M = 3$ covering a range of 55 km. It came into service in 1973 and was decommissioned in 2012. In 2013, the Meteor (MBDA UK, Fr) solid-fueled ramjet became also operative in some of the NATO countries. In Norway, the Nammo company has developed a solid fuel (HTPB) ramjet with metallic additives providing an acceleration up to Mach 3 in 50s. Russia (NPO) and India (DRDO) are developing a new solid-propellant SFDR (Solid Fuel Ducted Ramjet) version of the BrahMos supersonic cruise missile: this will be faster at roughly Mach 3.5 and 1.5 times smaller so that it could also be transported on a plane like the Mikoyan MiG-35 Fulcrum-F. Also Chine, in the last 10 years, developed the HD-1 cruise missile by the Hong Kong Blasting Company (GHBC) based in Hong Kong and the PL-12D, YJ-12 developed by the CASC (Research Institute of the China Aerospace Science and Technology Corporation).

Boeing is currently working on a hypersonic jet that could fly between Los Angeles and Tokyo in about three hours. Further, Boeing's HorizonX, through an agreement with the British Reaction Engines, is financing the development of the SABRE (Synergetic Air-Breathing Rocket Engine) engine, which is able to travel to Mach 5 in ramjet mode for supersonic flight and Mach 25 in rocket mode for hypersonic flight (Skylon). Tests Flight tests could start as early as 2025 for Skylon, and 2020 for the Saber. Small private companies as BOOM are also announced a 45 business class passengers for supersonic traffic between New York and London in 3 h and 25 min with service entry by 2025. This means the Boom Super Sonic Transport shall be 10% faster than the Concorde.

The European projects LAPCAT/LAPCAT2, HEXAFLY, STRATOFLY [8, 9], in which industries, research centers and universities are involved, investigated the feasibility of a commercial aircraft transport for 300 passengers flying at Mach $=$ 7–8, able to cover the distance Brussels-Sydney in about 2 h.

Dual mode Ramjet (ramjet + scramjet) for spaceflight applications has also been studied for both two stage to orbit and single stage to orbit systems [10].

References

1. R.S. Fry, A century of ramjet propulsion technology evolution. J. Propuls. Power **20**(1), 27–58, January-February (2004)
2. P.J. Waltrup, M.E. White, F. Zarlingo, E.S. Gravlin, History of Ramjet and Scramjet propulsion development for U.S. navy missiles. Johns Hopkins APL Tech. Digest **18**, 234–243 (1997)
3. P.W. Hewitt, B. Waltz, S. Vandiviere, Ramjet Tactical Missile Propulsion Status, AIAA 2002 Missile Sciences Conference [Classified and Unclassified Documents] 5–7 Nov 2002
4. F.T. Esenwein, C.L. Pratt, Design and performance of an integral Ramjet/Rocket, in *Proceedings of 2nd International Symposium on Air Breathing Engines*, Sheffield, England (Apr 1974)
5. H.P. Jenkins, Jr., Solid-fuel Ramjet developments by U.S. naval ordnance tests station, China Lake, California, NOTS TP 2177 (NAVORD Report 6459) (18 Nov 1958)
6. C.J. Crowell, G.C. Googins, Final report on development of solid fuel for Ramjet Sustainer Motor, U.S. Naval Missile Center Technical Memorandum No. NMC-TM-66–7 (4 May 1966)
7. "Ramjet Research Payoffs Believed Near. Aviat. Week and Space Technol. **105**, 71–77 (13 Sep 1976)
8. A. Ingenito, C. Bruno, S. Gulli, Preliminary sizing of hypersonic airbreathing airliner. Trans. Japan Soc. Aeronaut. Space Sci. Space Technol. **8**, No. ISTS27, 19–28 (2010), Japan. 1884-0485, ISSN: 1347-3840
9. A. Ingenito, S. Gulli, C. Bruno, G. Colemann, B. Chudoba, P.A. Czysz, Sizing of a fully integrated hypersonic commercial airliner. J. Aircr. **48**(6), 2161–2164 (2011). https://doi.org/10.2514/1.C000205
10. P.J. Waltrup, The Dual Combustor Ramjet: A Versatile Propulsion System for Tactical Missile Applications, AGARD CP-726, Paper No. 7 (May 1992)

Chapter 2
Fundamentals of Ramjet Engines

Abstract Ramjets are particular devices in which compression takes place by exploiting the kinetic energy of the flow. Through a system of oblique and normal shocks, the flow will be slowed down to subsonic speeds in the combustion chamber, ensuring suitable pressures for the flame anchoring. The flight altitude must guarantee an adequate inlet mass flow rate for the required thrust, and at the same time, it should avoid excessive pressure loads on the wing surfaces. A corridor of flight altitudes according to the Mach number is therefore defined to ensure adequate performance avoiding excessive dynamic loads.

The ramjet is an airbreathing engine that exploits the speed of the vehicle to compress and decelerate air entering the inlet by the ram effect in a fixed duct. Elements of the ramjet power cycle and flowpath from [1] are shown in Fig. 2.1. The air enters the intake where it is compressed and decelerated to subsonic speed at the combustor inlet. Within the combustor, fuel and air mix and burn in the presence of a flame holder that stabilizes the flame.

The combustor exhausts expand through the nozzle at speed greater than that of the entering air, generating a forward thrust.

Since the air compression is only due to the speed decrease through the inlet, an auxiliary engine needs to provide a forward velocity to start the cycle. Generally, booster rockets are externally or internally integrated to provide the initial Mach number to start the ramjet. Current ramjet air-to-air- or air-to-ground missiles use an integral rocket-ramjet configuration, which involves solid propellant in the aft combustion or mixing chamber to boost the system to ramjet-operating conditions. Once operative conditions are achieved, the nozzle and associated components are discarded, and the ramjet cycle begins.

Airbreathing over rocket propulsion shows higher performance at different Mach numbers, however, both ramjet and scramjet need an auxiliary engine to work (See Fig. 2.2).

Despite the lower Isp, hydrocarbon fuels are preferred where volumetric and operational constraints are critical, i.e., for hypersonic missile applications.

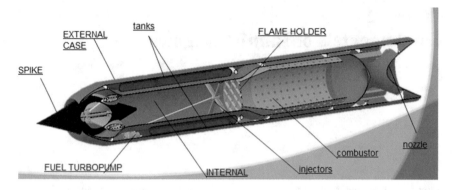

Fig. 2.1 Elements of the ramjet engine

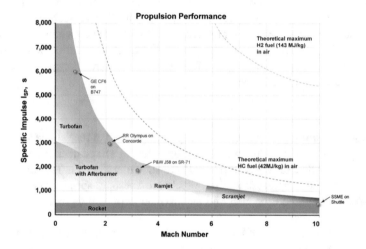

Fig. 2.2 Characteristic performance by engine type [4]

Airbreathing vehicles entail high dynamic pressure q in order to provide enough static pressures at the combustor inlet to stabilize the flame and to ensure the required thrust. To keep the dynamic pressure constant over the whole flight, at given ranges of Mach numbers correspond a range of altitudes: in fact, increasing the altitude, the density decreases, this means that to keep the dynamic pressure constant, the speed must increase. A typical Mach number–altitude airbreathing flight corridor [2, 3] is shown in Fig. 2.3. The upper boundary is characterized by higher altitudes and therefore, by lower ambient pressures and densities. Thermodynamic conditions in the ramjet combustor in terms of inlet pressure and temperature will determine the upper limit due to the low combustion efficiency.

At lower altitudes, the higher pressure loading and in turn, the higher skin temperatures define the lower boundary due to the material constraints. Increasing the Mach number, the total temperature increases, inducing dissociation, structure heating and

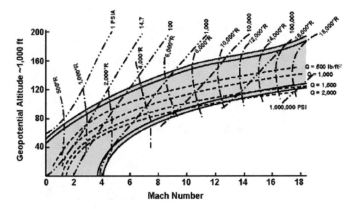

Fig. 2.3 Mach number–altitude airbreathing flight corridor [5]

distortion, and combustion inefficiency. At low Mach numbers, the static pressure and temperature at the combustor inlet are not enough for efficient combustion.

References

1. P.W. Hewitt, B. Waltz, S. Vandiviere, Ramjet Tactical Missile Propulsion Status, AIAA 2002 Missile Sciences Conference [Classified and Unclassified Documents] 5–7 November 2002
2. E.T. Curran, S. Murthy, *High-Speed Flight Propulsion Systems*, vol. 137 of Progress in Astronautics and Aeronautics (AIAA, 1991)
3. F.S. Billig, R.C. Orth, M. Lasky, Effects of thermal compression on the performance estimates of hypersonic Ramjets. J. Spacecr. Rockets **5**(9), 1076–1081 (1968)
4. https://upload.wikimedia.org/wikipedia/commons/4/4f/Specific-impulse-kk-20090105.png. consulted 10/05/2020
5. W.H. John, Flight Testing of Airbreathing Hypersonic Vehicles—NASA Technical Memorandum 4524 (1993)

Chapter 3
Design of Supersonic/Hypersonic Vehicles

Abstract Studies on supersonic/hypersonic vehicles by the NASA AMI-X program [1], date back to the '60s [2]. This program underlined that the approach of integrating individually optimized system elements (as routinely done in designing subsonic aircraft) yields a significant reduction in performance [3]. Vanderkerchove (VDK) and Czysz [4] identified a new approach, based on the Küchemann tau, to screen, identify, and visualize the design to first order over a range of feasible solutions.

3.1 Küchemann tau Figure of Merit

In supersonic/hypesonic vehicles, the design starts from its mission requirements [5, 6], i.e., range, nominal Mach number, payload and fuel: all those parameters are strictly correlated and all them depend on the vehicle shape. As an example, the vehicle surface area will increase L/D, this allowing a smaller amount of fuel required for a given dry weight. Decreasing the on-board fuel will also save the volume required to accommodate the fuel and consequently the weight of the structure. However, increasing the wings surface will also increase the structural weight and in turn the dry weight, requiring more fuel to supply the weight increase. This implies that there is a close competition and correlation between weights, performance, volumes and the shape of the vehicle.

Küchemann [7] introduced a shape factor tau, τ, defined as $\tau = \frac{V_{tot}}{S_{plan}^{1.5}}$ that identifies the figure of merit of a given vehicle. This parameter varies from tau $= 0$ for a flat plate and $\tau = 0.75225$ for a sphere (In fact, $\tau_{sphere} = \frac{V_{tot_{sphere}}}{S_{plan_{sphere}}^{1.5}}$, with: $V_{tot_{sphere}} = \frac{4}{3}\pi r^3$ and $S_{plan_{sphere}} = \pi r^2$, so $\tau_{sphere} = 0.75225$). Figure 3.1 shows the general trend with τ: as τ increases the Planform Surface decreases and the available volume increases.

In order to get the best solution, a set of equations, accounting for the shape, performance and weights must be simultaneously solved to achieve convergence solutions., Vanderkerchove (VDK) and Czysz [4] identified a new approach based on the simultaneous solution for the OWE (overall weight empty) and planform

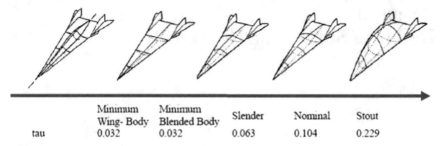

	Minimum Wing-Body	Minimum Blended Body	Slender	Nominal	Stout
tau	0.032	0.032	0.063	0.104	0.229

Fig. 3.1 General trends varying τ [7]

area (S_{plan}) equations, ensuring that the separately calculated available and required weights and total volumes (V_{tot}) converge for the given shape factor τ.

3.2   Breguet Range for Airbreathing Supersonic/Hypersonic Vehicles

The Breguet Range Equation relates the maximum range achievable with the fuel fraction, ff, the fuel combustion energy, Q_{cc}, the vehicle aerodynamics, L/D, and the propulsion efficiency, Isp.

For cruise conditions, the Breguet Range is:

$$\text{Breguet Range} = -RF \times \ln(1 - ff) \tag{3.1}$$

where the range factor, RF, is given by:

$$RF = \theta \times Q_{cc} \times \frac{L}{D} = a \times M \times I_{sp} \times \frac{L}{D} \tag{3.2}$$

The fuel fraction ff:

$$ff = \frac{W_{fuel}}{TOGW} \tag{3.3}$$

is defined as the ratio between the fuel weight, W_{fuel}, divided by the takeoff gross weight, TOGW that is the sum of the payload, the fuel, the various systems and the propellant weight:

$$TOGW = W_{pay} + W_{fuel} + W_{sys} + W_{prop}$$

The Range Factor equation (see Eq. 3.1) shows that there is a maximum range value that depends on the Mach number, the fuel configuration, and the propellant. In fact, increasing the Mach number will increase the range factor till a maximum value depending on the fuel. For hydrogen as fuel, above Mach 6 the Range Factor stays essentially constant at 1200 nm (see Fig. 3.2). Hydrocarbons (methane in red and kerosene in green) achieve a maximum speed of about $M = 8$ and a range factor about 2.5 times smaller than hydrogen.

Depending on the shape, different L/D may be achieved. Figure 3.3 shows the maximum experimental L/D (and consequently the maximum range achievable) for different configurations (Küchemann tau, $\tau = \frac{V_{\text{tot}}}{S_{\text{plan}}^{1.5}}$).

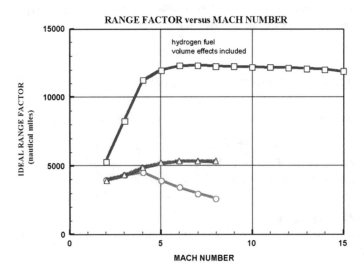

Fig. 3.2 RF versus Mach number [8]

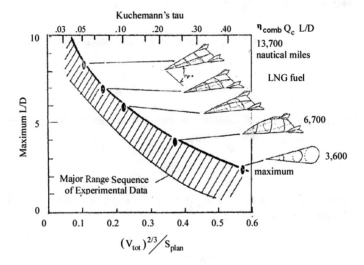

Fig. 3.3 Major Range sequence of Experimental L/D data [7]

For the same planform area, S_{plan}, as the total volume increases, V_{tot}, the aerodynamic drag increases, accordingly: therefore smaller τ maximize L/D. The infinitely thin flat plate ($\tau = 0$) would represent the most efficient hypersonic lifting surface, though, the flat plate cannot contain any volume for payload, engines, fuel and general devices. Therefore, the most promising vehicle architecture will be that with the minimum volume feasible to attain the mission requirements and the minimum planform area to ensure the highest L/D. In fact, increasing the planform area accordingly increases the structural weight and costs related to the structure. A representative value of tau for ramjet engine-powered missiles is ~0.19.

3.3 Hypersonic Convergence Equations

When sizing hypersonic vehicles, the set of unknowns variables are:

- **7** for the geometry: $\tau = \frac{V_{tot}}{S_{plan}^{1.5}}$, S_{plan}, S_{wet}, V_{tot}, V_{pay}, V_{void}, V_{fuel}
- **1** for performance: $L/D = f(Ma, \tau)$
- **5** for the weights: $TOGW$, W_{sys}, W_{prop}, W_{fuel}, W_{str}
- **1** for technology: $K_w(\tau)$

The unknowns are **14.**

To determine these unknowns, a set of equations must be identified and iterated until convergence.

3.3.1 Weights and Volumes Hypersonic Convergence

The minimum total volume, Vtot, required to accommodate the payload, the fuel, and onboard systems, i.e.,:

$$V_{tot} = V_{pay} + V_{fuel} + V_{void} \tag{3.4}$$

has to convergence with the volume calculated from the Küchemann tau shape parameter for a given planform area S_{plan}:

$$V_{tot} = \tau S_{plan}^{1.5}$$

The payload volume in Eq. 3.4 is given by the payload mass and the payload density, which are usually known:

$$V_{pay} = \frac{W_{pay}}{\rho_{pay}} \tag{3.5}$$

The void volume is the volume required to accommodate all systems:

$$V_{\text{void}} = V_{\text{tot}} \times (1 - \eta_V) \tag{3.6}$$

where η_V is assumed around 0.7.

The fuel volume depends on the fuel mass required for the mission range over the fuel storage density. Increasing the range factor, RF, will also increase the fuel mass fraction onboard, ff, i.e., the fuel mass over the take-off gross weight, TOGW and consequently the fuel volume:

$$ff = \frac{W_{\text{fuel}}}{\text{TOGW}} = 1 - e^{-\frac{\text{Range}}{I_{sp} \times \Delta V \times \frac{L}{D}}} \tag{3.7}$$

In order to calculate the TOGW, the weight of the payload, W_{pay}, the fuel, W_{fuel}, all devices, W_{sys} and the propulsion engine, W_{prop} have to be calculated:

$$\text{TOGW} = W_{\text{pay}} + W_{\text{fuel}} + W_{\text{sys}} + W_{\text{prop}} \tag{3.8}$$

The systems' weight is calculated from:

$$W_{\text{sys}} = \frac{W_{\text{sys}}}{\text{TOGW}} \text{TOGW} = r_{\text{sys}} \times \text{TOGW} \tag{3.9}$$

where r_{sys} is usually assumed to be 0.07.

The engines' weight is estimated by assuming:

$$W_{\text{prop}} = \frac{W_{\text{prop}}}{\text{Thrust}} \left(\frac{1}{L/D} \right) \text{TOGW} = \frac{1}{\text{ETW}} \left(\frac{1}{L/D} \right) \text{TOGW} \tag{3.10}$$

where ETW is the engine thrust-to-weight, here estimated 8.3 for ramjet engines.

The structural weight depends on the wetted surface and the structural weight index:

$$W_{\text{str}} = \frac{W_{\text{str}}}{S_{\text{wet}}} \frac{S_{\text{wet}}}{S_{\text{plan}}} S_{\text{plan}} = I_{\text{str}} \times K_w \times S_{\text{plan}} \tag{3.11}$$

where K_w is a parameter relating the planform area to the wetted area:

$$S_{\text{wet}} = K_w(\tau) \times S_{\text{plan}} \tag{3.12}$$

This parameter is function of tau and, for a blended body configuration, is:

$$K_w = -62.217 \times \tau^3 + 29.904 \times \tau^2 - 1.581 \times \tau + 2.469 \tag{3.13}$$

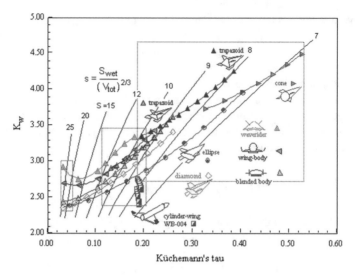

Fig. 3.4 Kw versus tau for different vehicle configurations [8]

Figure 3.4 shows that different relationships hold for different vehicle configurations.

The lower the structural weight index and the more advanced technological is the solution required. Three I_{str} have been assumed to calculate the structural weight: 15, 18, 21. The I_{str} defines the level of maturity and technology of structural materials (see Fig. 3.5).

Kw increases with tau, and as a consequence, the wetted area should increase with tau and hence the friction drag. However, Eq. 2.12 shows that Swet depends on the contribution of K_w and S_{plan}, and the latter decreases with tau (for instance, from a waverider, with the minimum tau, to a sphere, with the highest tau), therefore also the wetted surface decreases (see Fig. 3.6) with tau. The platform area decrease lowers the attainable lift, and therefore L/D decreases with tau.

In fact, L/D is a function of tau and of the Mach number [9]:

$$\frac{L}{D} = \frac{A(M + B)}{M} \left[\frac{1.0128 - 0.2797 \cdot \ln(\tau/0.03)}{1 - \frac{M^2}{673}} \right] \tag{3.14}$$

where A = 6 and B = 2.

Figure 3.7 shows L/D decreasing with the increase of either the Mach number or tau.

Concluding, the equations to be solved are:

6 for geometric unknowns (these relate the vehicle architecture to the vehicle performance and mission requirements by means of the Küchemann tau):

$$V_{tot} = \tau \times S_{plan}^{1.5} \tag{3.15}$$

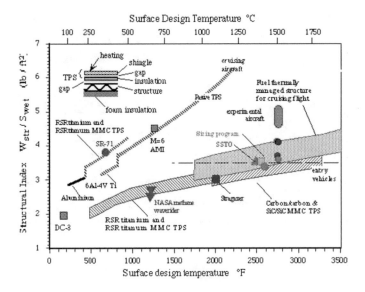

Fig. 3.5 I_{str} versus surface temperature [8]

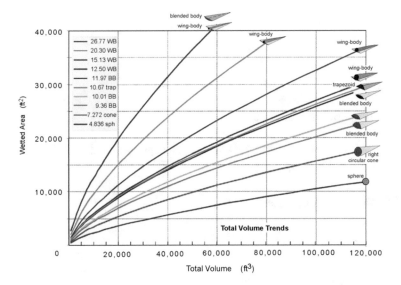

Fig. 3.6 Wetted area versus Total Volume [7]

$$S_{\text{wet}} = K_w(\tau) \times S_{\text{plan}} \tag{3.16}$$

$$V_{\text{fuel}} = W_{\text{fuel}}/\rho_{\text{fuel}} \tag{3.17}$$

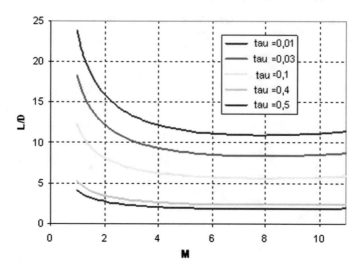

Fig. 3.7 L/D versus M with tau as parameter

$$V_{\text{pay}} = W_{\text{pay}}/\rho_{\text{pay}} \tag{3.18}$$

$$V_{\text{void}} = V_{\text{tot}} \times (1 - \eta_V) \tag{3.19}$$

$$V_{\text{tot}} = V_{\text{pay}} + V_{\text{fuel}} + V_{\text{void}} \tag{3.20}$$

1 for the performance variable L/D:

$$\frac{L}{D} = \frac{A(M + B)}{M} \left[\frac{1.0128 - 0.2797 \cdot \ln(\tau/0.03)}{1 - \frac{M^2}{673}} \right] \begin{array}{l} A = 6 \\ B = 2 \end{array} \tag{3.21}$$

1 for the technology variable K_w:

$$K_w = \frac{S_{\text{wet}}}{S_{\text{plan}}} = \tau \cdot e^{1.414 - 1.415\ln(\tau) - 0.731\ln^2(\tau) - 0.272\ln^3(\tau) - 0.031\ln^4(\tau)} \tag{3.22}$$

And **5** equations for the weights:

$$\text{TOGW} = W_{\text{pay}} + W_{\text{fuel}} + W_{\text{sys}} + W_{\text{prop}} \tag{3.23}$$

$$W_{\text{fuel}} = \left(1 - e^{-\frac{\text{Range}}{I_{sp} \times \Delta V \times \frac{L}{D}}} \right) \text{TOGW} \tag{3.24}$$

$$W_{\text{sys}} = \frac{W_{\text{sys}}}{\text{TOGW}} \text{TOGW} = r_{\text{sys}} \times \text{TOGW} \tag{3.25}$$

$$W_{\text{prop}} = \frac{W_{\text{prop}}}{\text{Thrust}} \left(\frac{1}{L/D}\right) \text{TOGW} = \frac{1}{ETW} \left(\frac{1}{L/D}\right) \text{TOGW} \tag{3.26}$$

$$W_{\text{str}} = \frac{W_{\text{str}}}{S_{\text{wet}}} \frac{S_{\text{wet}}}{S_{\text{plan}}} S_{\text{plan}} = I_{\text{str}} \times K_w \times S_{\text{plan}} \tag{3.27}$$

1 for the fuel volume:

$$V_{\text{fuel}} = W_{\text{fuel}}/\rho_{\text{fuel}} \tag{3.28}$$

This set of equations must be solved to get the best performing vehicle's volumes and weights. In fact, once chosen the Mach number, the Range, the flight Altitude, the fuel, the payload weight, the ETW (8.5 for ramjet engines), r_{sys}, η_v, I_{str} and τ, the sizing process identifies all the unknowns, and in particular the thrust, the propellant weight and the inlet area. This is the starting point for the ramjet engine sizing.

References

1. I.M. Blankson, J.S. Pyle, NASA's Hypersonic Flight Research Program, AIAA 93-0308, 31st ASM, Jan. 11–14, Reno, NV (1993)
2. R.S. Fry, A century of Ramjet propulsion technology evolution. J. Propuls. Power **20**(1), 27–58, January-February (2004)
3. E.T. Curran, Introduction, in *High-Speed Flight Propulsion Systems*, ed. by S.N.B. Murthy and E.T. Curran (AIAA, Reston, VA, 1991), pp. 1–20
4. P.A. Czysz, C. Bruno, *Future Spacecraft Propulsion Systems*, 2nd ed., Chapter 5 (Springer-Praxis, London, 2009)
5. A. Ingenito, C. Bruno, S. Gulli, Preliminary sizing of hypersonic airbreathing airliner, in *Transactions of the Japan Society for Aeronautical and Space Sciences, Space Technology*, Vol. 8, No. ISTS27, pp. 19–28 (2010), Japan. 1884-0485, ISSN: 1347-3840
6. A. Ingenito, S. Gulli, C. Bruno, G. Colemann, B. Chudoba, P.A. Czysz, Sizing of a fully integrated hypersonic commercial airliner. J. Aircraft **48**(6), 2161–2164 (2011). https://doi.org/10.2514/1.c000205
7. D. Koelle, Sänger Advanced Space Transportation System - Progress Report 1990, AIAA-90-5200 (1990)
8. C. Bruno, P.A. Czysz, *Future Spacecraft Propulsion Systems: Enabling Technologies for Space Exploration*. Springer Praxis Books, 16 Mar 2009
9. W. Heiser, D. Pratt, D. Daley, U. Mehta, *Hypersonic Airbreathing Propulsion* (AIAA, 1994)

Chapter 4
Ramjet Engines Performance

Abstract Ramjet performance is calculated by means of the well-acknowledged Bryton cycle. The real Bryton cycle accounting for friction losses, pressure recovery, combustion efficiency and the thermal chocking, has been analyzed to establish the greatest operative conditions for the ramjet engine.

4.1 Ideal Ramjet Cycle

The ideal ramjet cycle is well acknowledged as the Brayton cycle. Air enters in 0 at the flight velocity M0. The flow is decelerated through a convergent–divergent diffusor at subsonic speeds ($M_2 \ll 1$) in the combustion chamber: the total pressure is ideally assumed constant from the inlet to the outlet of the ramjet. Decreasing the dynamic pressure, the static pressure increases. Station 0 is the entrance of the external inlet and station 1 is the entrance of the internal inlet. Downstream of the isobaric combustion (between stations 2 and 7), exhausts are expanded through the de Laval nozzle increasing the dynamic pressure and reducing the static pressure (Fig. 4.1).

The equivalent T-S diagram of the Brayton cycle is defined in Fig. 4.2.

A short description of the ideal cycle follows.

Stations 0–2

In the ideal cycle, the compression is assumed to be isentropic. On the T-S diagram, this corresponds to the segment 1–02. Upstream of the inlet, at station 0, it is possible to define:

$$\frac{T_{t0}}{T_0} = 1 + \frac{\gamma - 1}{2}M_0^2 = \theta_0 \tag{4.1}$$

$$\frac{p_{t0}}{p_0} = \left(1 + \frac{\gamma - 1}{2}M_0^2\right)^{\frac{\gamma}{\gamma-1}} = \delta_0 \tag{4.2}$$

© The Author(s), under exclusive license to Springer Nature Switzerland AG 2021
A. Ingenito, *Subsonic Combustion Ramjet Design*,
SpringerBriefs in Applied Sciences and Technology,
https://doi.org/10.1007/978-3-030-66881-5_4

Fig. 4.1 Sketch of the
ramjet cycle

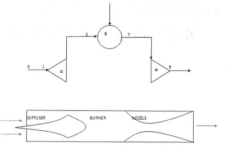

Fig. 4.2 TS Brayton cycle

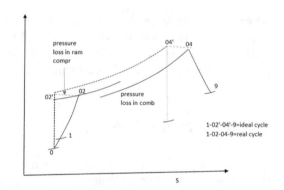

$$\delta_0 = \theta_0^{\frac{\gamma}{\gamma-1}} \tag{4.3}$$

$$\theta_0 = \delta_0^{\frac{\gamma-1}{\gamma}} \tag{4.4}$$

Since the process is assumed isentropic, the total temperature and total pressure
stay constant:

$$\frac{T_{t0}}{T_{t2}} = 1 \tag{4.5}$$

$$\frac{p_{t0}}{p_{t2}} = 1 \tag{4.6}$$

A more realistic compression would replace the isentropic by polytropic compression, with an air exponent of order 1.23 to 1.3.

Stations 2–7

Assuming M_2 as the Mach number at the combustor inlet (typical values are from 0.1 to 0.4), temperature, pressure and velocity are:

$$T_2 = \frac{T_{t0}}{1 + \frac{\gamma-1}{2}M_2^2} \tag{4.7}$$

$$p_2 = \frac{p_{t0}}{\left(1 + \frac{\gamma-1}{2}M_2^2\right)^{\frac{\gamma}{\gamma-1}}} \tag{4.8}$$

$$U_2 = \sqrt{\gamma RT_2} M_2 \tag{4.9}$$

Being the combustion isobaric, i.e., constant static pressure in the combustor, even though the total temperature increases due to the combustion heat release, the stagnation pressure holds constant:

$$\frac{p_{t9}}{p_{t2}} = 1 \tag{4.10}$$

The adiabatic flame temperature is calculated by the first law of thermodynamics:

$$du + pdv = dh - vdp = \delta q + \delta wR \tag{4.11}$$

where δq is the heat transfer from the surroundings, δwR is the frictional work, du is the change of internal energy and pdv is the work due to volumetric changes, and $v = 1/\rho$ is the specific volume.

The specific enthalpy h of an ideal gas is related to the specific internal energy u by

$$h = u + pv \tag{4.12}$$

For an ideal gas, it is possible to write:

$$h = u + RT/W \tag{4.13}$$

where W is the molecular weight of the species.

In a multicomponent system, the specific internal energy and specific enthalpy are the mass weighted sums of the specific quantities of all species

$$u = \sum_{i=1}^{k} Y_i u_i; \ h = \sum_{i=1}^{k} Y_i h_i$$

For an ideal gas, the partial specific enthalpy is related to the partial specific internal energy by:

$$h_i = u_i + \frac{RT}{W_i}, i = 1, 2, \ldots, k \tag{4.14}$$

and both depend only on temperature:

$$h_i = h_{i,\text{ref}} + \int_{T_{\text{ref}}}^{T} c_{pi} dT i = 1, 2, \ldots, k \tag{4.15}$$

where c_{pi} is the specific heat capacity at constant pressure and $h_{i,\text{ref}}$, is the reference enthalpy at the reference temperature $T_{\text{ref}} = 298.15\ K$.

C_{pi}, H_i and S_i may be calculated from the NASA polynomials [1], where the coefficients depend on the specific gas:

$$\frac{C_{pi}}{R} = a_1 + a_2 T/K + a_3(T/K)^2 + a_4(T/K)^3 + a_5(T/K)^4$$

$$\frac{H_i}{RT} = a_1 + a_2 \frac{T/K}{2} + a_3 \frac{(T/K)^2}{3} + a_4 \frac{(T/K)^3}{4} - a_5 \frac{(T/K)^4}{5} + \frac{a_6}{T/K}$$

$$\frac{S_i}{R} = a_1 \ln(T/K) + a_2 T/K + a_3 \frac{(T/K)^2}{2} + a_4 \frac{(T/K)^3}{3}$$
$$+ a_5 \frac{(T/K)^4}{4} + a_7 + \ln\left(\frac{p}{p_0}\right)$$

For an adiabatic system ($\delta q = 0$) at constant pressure (dp = 0) and neglecting the work done by friction ($\delta w R = 0$):

$$dh = 0 \tag{4.16}$$

Integrating from the unburnt to the burnt state, it follows:

$$h_u = h_b \tag{4.17}$$

or

$$\sum_{i=1}^{k} Y_{i,u} h_{i,u} = \sum_{i=1}^{k} Y_{i,b} h_{i,b} \tag{4.18}$$

By combining with Eq. 3.15, the Eq. 3.17 reads:

$$\sum_{i=1}^{k}(Y_{i,u} - Y_{i,b})h_{i,\text{ref}} = \int_{T_{\text{ref}}}^{T_b} c_{p,b}dT - \int_{T_{\text{ref}}}^{T_u} c_{p,u}dT \tag{4.19}$$

where cp, b and cp, u are those of the mixture, and are calculated with the mass fractions of the burnt and unburnt gases:

$$c_{p,b} = \sum_{i=1}^{k} Y_{i,b}c_{pi}(T); \; c_{p,u} = \sum_{i=1}^{k} Y_{i,u}c_{pi}(T)$$

A simplified methodology to estimate the flame temperature consists in the assumption of the fast chemistry, i.e., a single overall step where air and fuel become products.

For such one-step global reaction, burned and unburned fuel and oxidizer are calculated by [2]:

$$Y_{i,u} - Y_{i,b} = (Y_{F,u} - Y_{F,b})\frac{v_i W_i}{v_F W_F} i = 1, 2, \ldots, k \tag{4.20}$$

$$Y_{i,u} - Y_{i,b} = (Y_{O_2,u} - Y_{O_2,b})\frac{v_i W_i}{v_{O_2} W_{O_2}} i = 1, 2, \ldots, k \tag{4.21}$$

and consequently:

$$\sum_{i=1}^{k}(Y_{i,u} - Y_{i,b})h_{i,\text{ref}} = \frac{(Y_{F,u} - Y_{F,b})}{v_F W_F} \sum_{i=1}^{k} v_i W_i h_{i,\text{ref}} \tag{4.22}$$

The heat of combustion is given by:

$$Q = -\sum_{i=1}^{k} v_i W_i h_{i,\text{ref}} = -\sum_{i=1}^{k} v_i H_i \tag{4.23}$$

Since the heat of combustion changes very little with temperature, it is possible to write:

$$Q_{\text{ref}} = -\sum_{i=1}^{k} v_i H_{i,\text{ref}} \tag{4.24}$$

Assuming c_p as an average constant value and $Q = Q_{\text{ref}}$, the adiabatic flame temperature for a rich mixture ($Y_{O2,b} = 0$) is:

$$T_b - T_u = \frac{Q_{ref} Y_{O_2,u}}{c_p \nu'_{O_2} W_{O_2}} \tag{4.25}$$

and for lean mixtures ($Y_{F,b} = 0$) is:

$$T_b - T_u = \frac{Q_{ref} Y_{F,u}}{c_p \nu'_F W_F} \tag{4.26}$$

Assuming Z as the mixture fraction, ($0 < Z = Yf/Yox < 1$), the temperature of the unburnt mixture for a given Z is calculated by:

$$T_u(Z) = T_2 - Z(T_2 - T_1) \tag{4.27}$$

where T_2 is the temperature of the oxidizer stream and T_1 that of the fuel. Therefore:

$$T_b(Z) = T_u(Z) + \frac{Q_{ref} Y_{F,1}}{c_p \nu'_F W_F} ZZ \le Z_{st} \tag{4.28}$$

$$T_b(Z) = T_u(Z) + \frac{Q_{ref} Y_{O_2,2}}{c_p \nu'_{O_2} W_{O_2}} (1 - Z)Z \ge Z_{st} \tag{4.29}$$

At $Z = Z_{st}$, the maximum flame temperature is given by:

$$T_{st} = T_u(Z_{st}) + \frac{Y_{F,1} Z_{st} Q_{ref}}{c_p \nu'_F W_F} = T_u(Z_{st}) + \frac{Y_{O_2,2}(1 - Z_{st}) Q_{ref}}{c_p \nu'_{O_2} W_{O_2}} \tag{4.30}$$

Figure 4.3 shows the temperature of the mixture for different mixture fractions. Having made the constant c_p assumption, the temperature change with Z is necessarily a straight line.

For the combustion of a pure fuel ($Y_{F,1} = 1$) in air ($Y_{O2,2} = 0.232$) with $T_{u,st} = 300$ K values for T_{st} are given in Table 4.1 using $c_p = 1.4$ kJ/kg/K.

Fig. 4.3 T versus Z_st

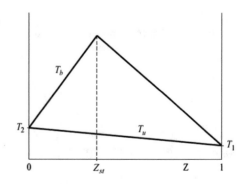

Table 4.1 Stoichiometric mixture fractions and stoichiometric flame temperatures for some hydrocarbon–air mixtures

Fuel	Z_{st}	T_{st} [K]
CH_4	0.05496	2263.3
C_2H_6	0.05864	2288.8
C_2H_4	0.06349	2438.5
C_2H_2	0.07021	2686.7
C_3H_8	0.06010	2289.7

As said, the assumption of one-step complete combustion is an approximation because it disregards the possibility of dissociation of combustion products. In order to account for these effects, Fig. 14 shows the adiabatic temperature over equivalence ratio calculated with the CEA 600 code for different aviation fuels, assuming chemical equilibrium rather than complete combustion (Fig. 4.4).

The temperature ratio T_{04}/T_{02} between the combustor outlet and inlet is shown in Fig. 15: it varies in the range of 2–6 (Fig. 4.5).

Stations 4–9

The ideal ramjet engine cycle, the expansion processes is assumed to be isentropic so that the stagnation pressure keeps constant throughout the nozzle:

$$p_{t0} = p_{t9} \tag{4.31}$$

$$\frac{p_{t0}}{p_0} = \left(1 + \frac{\gamma - 1}{2} M_0^2\right)^{\gamma/\gamma - 1} \tag{4.32}$$

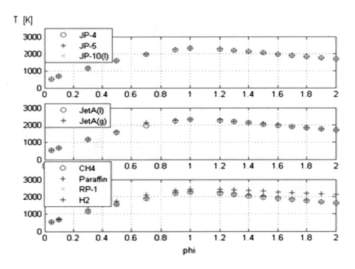

Fig. 4.4 T versus ϕ, $h = 15\,$km $M_0 = 2$; $T_2 = 390.6\,$K: $p_2 = 94175\,$Pa

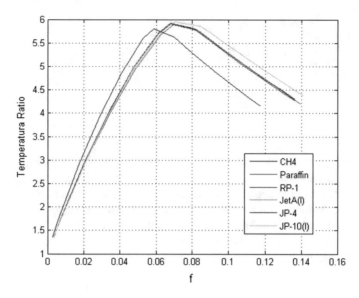

Fig. 4.5 T ratio versus ϕ, $M_0 = 2$; $h = 15$ km, $T_2 = 390.6$ K: $p_2 = 94175$ Pa

$$\frac{p_{t9}}{p_9} = \frac{p_{t0}}{p_0} = \left(1 + \frac{\gamma - 1}{2} M_e^2\right)^{\gamma/\gamma - 1} \tag{4.33}$$

where M_0 is the flight Mach number and M_e is the exit plane Mach number. Assuming ideal expansion and matching pressures at the nozzle exit ($p_e = p_0$) :

$$\frac{p_{t0}}{p_0} = \frac{p_{t9}}{p_e} \tag{4.34}$$

$$M_e = M_0 \tag{4.35}$$

This implies that:

$$\frac{U_e}{U_0} = \frac{a_e}{a_0} = \frac{\sqrt{T_e}}{\sqrt{T_0}} = \sqrt{\frac{T_{t9}}{T_{t0}}} = \sqrt{\frac{T_{t4}}{T_{t3}}} \tag{4.36}$$

Therefore the static temperature at the nozzle exit, T_9, is:

$$T_9 = \frac{T_{t9}}{1 + \frac{\gamma - 1}{2} M_9^2} \tag{4.37}$$

4.2 Real Ramjet Cycle

In a real ramjet cycle, although assuming no friction losses, the combustion process within a constant section channel decreases the stagnation pressure. Approximating the flow as one dimensional, i.e., assuming uniform temperature and velocity through the flow direction and the friction is negligible, the momentum equation applied to the combustor is:

$$(p_3 - p_4)A - D = \dot{m}_4 u_4 - \dot{m}_3 u_3 \tag{4.38}$$

where D is the resistance (drag) exerted on the flow, in the opposite direction, by the flame holders. Drag is proportional to the dynamic pressure input ($\frac{1}{2}\rho u_3^2$) by a factor K representing the ratio between the pressure drop (caused by friction) and the upstream dynamic pressure. The station 3 stands for the flow conditions in the combustion chamber without accounting for the Rayleigh's effect. Station 4 characterizes the flow at the combustor exit.

$$p_3 - p_4 = \rho_4 u_4^2 - \rho_3 u_3^2 + K\left(\frac{1}{2}\rho_3 u_3^2\right) \tag{4.39}$$

K assumes the value of 1 or 2 if walls friction is accounted or not.
Assuming for simplicity $\gamma_3 = \gamma_4 = \gamma$, it follows:

$$\frac{p_3}{p_4} = 1 + \gamma M_4^2 - \gamma M_3^2\left(\frac{p_3}{p_4}\right) + K\frac{\gamma M_3^2}{2}\frac{p_3}{p_4} \tag{4.40}$$

Or

$$\frac{p_3}{p_4} = \frac{1 + \gamma M_4^2}{1 + \gamma M_3^2\left(1 - \frac{K}{2}\right)} \tag{4.41}$$

Consequently, the ratio between the total pressure at the combustor inlet and outlet is:

$$\frac{p_{04}}{p_{03}} = \frac{1 + \gamma M_3^2\left(1 - \frac{K}{2}\right)}{1 + \gamma M_4^2}\left[\frac{1 + \frac{\gamma-1}{2}M_4^2}{1 + \frac{\gamma-1}{2}M_3^2}\right]^{\frac{\gamma}{\gamma-1}} \tag{4.42}$$

Assuming the mass flow rate of fuel negligible with respect of air, a good approximation is $\dot{m}_4 = \dot{m}_3$.

where

$$\rho_4 u_4 = \rho_3 u_3 \tag{4.43}$$

Being $p = \rho R T$, the previous equation becomes:

$$\frac{p_3}{p_4} = \frac{u_4}{u_3}\frac{T_3}{T_4} = \frac{M_4}{M_3}\sqrt{\frac{T_3}{T_4}} \tag{4.44}$$

Now, since the total temperatures are known (i.e., are calculated by combustion relationships):

$$\frac{T_3}{T_4} = \frac{T_{03}}{T_{04}}\frac{\left(1 + \frac{\gamma-1}{2}M_4^2\right)}{\left(1 + \frac{\gamma-1}{2}M_3^2\right)} \tag{4.45}$$

and combining Eq. 3.44 with Eq. 3.45, it follows:

$$\frac{p_3}{p_4} = \frac{M_4}{M_3}\sqrt{\frac{T_{03}}{T_{04}}\frac{\left(1 + \frac{\gamma-1}{2}M_4^2\right)}{\left(1 + \frac{\gamma-1}{2}M_3^2\right)}} \tag{4.46}$$

Finally, from Eq. 3.41 and Eq. 3.46, it follows:

$$\frac{T_{04}}{T_{03}} = \frac{M_4^2\left(1 + \frac{\gamma-1}{2}M_4^2\right)}{M_3^2\left(1 + \frac{\gamma-1}{2}M_3^2\right)}\frac{\left[1 + \gamma M_3^2\left(1 - \frac{K}{2}\right)\right]^2}{\left[1 + \gamma M_4^2\right]^2} \tag{4.47}$$

This equation shows that if $K = 0$ (no losses), combustion is described by a Rayleigh flow.

If $K = 1$, losses are due only to the flameholder and for $K = 2$ also frictional losses are accounted.

For $K > 0$ and $T_{04}/T_{03} > 1$, the Mach number increases toward the limit value (when thermal choking takes place) $M_4 = 1$.

For example, assuming:

- $0.2 < M_3 < 0.6$
- $M_4 = 1$
- $K = 1$ or $K = 2$

The maximum T_{04}/T_{03} that can be supplied to the flow before the thermal choking decreases by increasing the Mach number at the combustor inlet (see Figs. 4.6 and 4.7).

Assuming, for example, a Mach number at inlet of 0.2, the maximum temperature ratio is 5. Figure 4.5 shows that to keep the temperatures ratios below 5, fuel fraction smaller than 0.5 is mandatory.

Equation 4.45 shows that heating and fiction losses within the combustor decrease the total pressure (see Fig. 4.8).

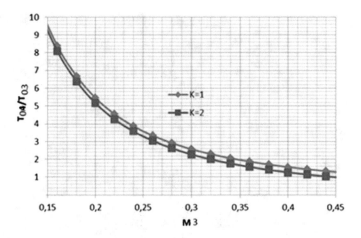

Fig. 4.6 T_{04}/T_{03} versus M_3 assuming $M_4 = 1$

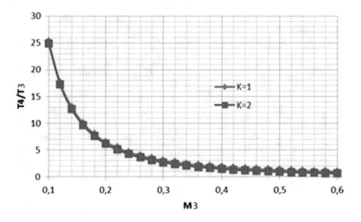

Fig. 4.7 T_4/T_3 versus M_3 assuming $M_4 = 1$

Let us define for these total pressure losses, a combustor "pneumatic" efficiency:

$$\eta_{pc} = \frac{p_{09r}}{p_{t0}} \tag{4.48}$$

If total pressure does not keep constant, the Mach number at the exit may be expressed as follows:

$$M_{9r}^2 = \frac{2}{\gamma - 1}\left[\left(1 + \frac{\gamma - 1}{2}M_0^2\right)\left(\eta_c\frac{p_0}{p_9}\right)^{\frac{\gamma-1}{\gamma}} - 1\right] \tag{4.49}$$

Assuming an adapted nozzle, i.e., $p_0 = p_9$.

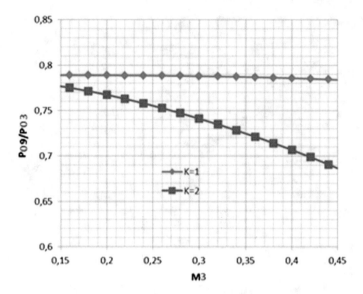

Fig. 4.8 p_{04}/p_{03} versus M_3

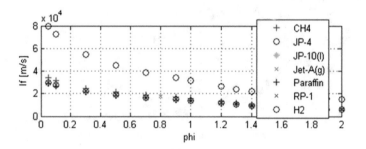

Fig. 4.9 I_{sp} versus Φ, $M_0 = 2$, $h = 15\,\text{km}$, $T_2 = 390.6\,\text{K}$: $p_2 = 94175\,\text{Pa}$

$$M_{9r}^2 = \frac{2}{\gamma - 1}\left[\eta_c^{\frac{\gamma-1}{\gamma}}\left(1 + \frac{\gamma - 1}{2}M_0^2\right) - 1\right] \tag{4.50}$$

and:

$$U_e = \sqrt{\gamma R T_{9r}}\,M_{9r} \tag{4.51}$$

Hence, the temperature at the nozzle exit is:

$$T_{9r} = \frac{T_{t9r}}{1 + \frac{\gamma-1}{2}M_{9r}^2} \tag{4.52}$$

4.3 Thrust Parameters and Efficiencies

The ideal theoretical thrust is purely calculated by:

$$T = \dot{m}_e U_e - \dot{m}_0 U_0 + (p_e - p_0) A_e \tag{4.53}$$

where

$$\dot{m}_e = (1 + f)\dot{m}_0 \tag{4.54}$$

$$f = \frac{\dot{m}_f}{\dot{m}_0} \tag{4.55}$$

Equation 3.55 may be written in convenient dimensionless form:

$$\frac{T}{\dot{m}_0 U_0} = (1 + f)\frac{U_e}{U_0} - 1 + \frac{p_0 A_e}{\dot{m}_0 U_0}\left(\frac{p_e}{p_0} - 1\right) \tag{4.56}$$

Assuming again $p_e = p_o$, this becomes:

$$\frac{T}{\dot{m}_0 a_0} = M_0\left[(1 + f)\frac{U_e}{U_0} - 1\right] \tag{4.57}$$

where $M_0 = U_0/a_0$, and a_0 is the local speed of sound.

$$\frac{T}{\dot{m}_0 a_0} = M_0\left[(1 + f)\sqrt{\frac{T_{t9}}{T_{t2}}} - 1\right] \tag{4.58}$$

Equation 4.60 highlights some interesting features of the ramjet engine:

1. ramjets do not develop static thrust: that is why they must be moved to deliver the thrust.
2. this device relies on "ram" compression of the air and has no moving parts (no spinning compressor to compress the air prior to combustion). Consequently, high flight speeds are required to get an efficient compression of the air.
3. performance relies on the stagnation temperature rise across the burner.

Let us introduce the specific impulse as the ratio between the thrust and the fuel mass flow rate:

$$I_{sp} = \frac{T}{\dot{m}_f} \tag{4.59}$$

and the specific thrust as:

$$I_a = \frac{T}{\dot{m}_o} [m/s] \tag{4.60}$$

The thrust specific fuel consumption is:

$$TSFC = \frac{\dot{m}_f}{T} [kg/h/N] \tag{4.61}$$

Figures 4.9 and 4.10 show respectively the specific impulse, I_{sp}, and the specific thrust, I_a, (calculated by the CEA 600 code), as function of the equivalence ratio and for different fuels.

Figures 4.10 and 4.11 show the specific thrust I_a, as a function of the equivalence ratio, respectively for different fuels and flight Mach numbers (assuming JP4 as fuel). Figure 22 shows the specific thrust I_a, for JP4 as fuel, as a function of the flight Mach number, assuming three different values for the flame temperature. These maximum temperatures are achievable assuming the equivalence ratios in Fig. 23. The $TSFC$ values are in the range of 0.15–0.4, its minimum is around $M = 3$. Assuming a $TSFC$ of 0.15, in order to produce 100 kg of thrust, 150 kg/h of JP4 should be burned (Figs. 4.12, 4.13 and 4.14).

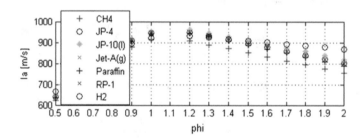

Fig. 4.10 I_a versus Φ, $M_0 = 2$; $h = 15$ km, $T_2 = 390.6$ K: $p_2 = 94175$ Pa

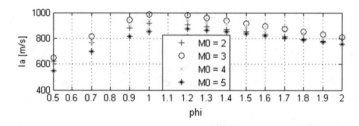

Fig. 4.11 I_a versus Φ for JP4 at different M_0, $h = 15$ km; $T_a = 217$ K

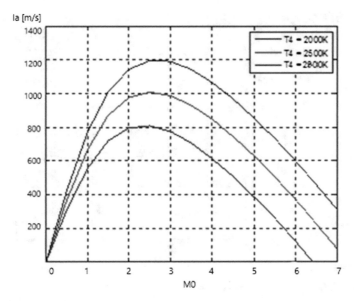

Fig. 4.12 I_a versus M_0 at $\Phi = 1$

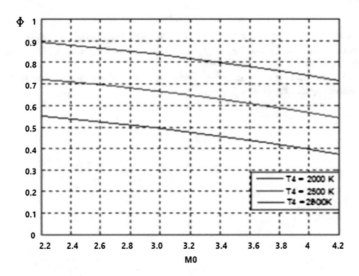

Fig. 4.13 Φ versus M_0

The analysis of other candidate fuels for ramjet applications in terms of proprieties and performance is shown in the next section.

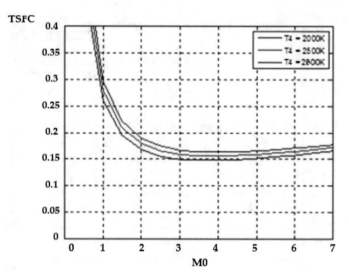

Fig. 4.14 TSFC of JP- 4 versus M and T_{max}

References

1. B. McBride, S. Gordon, J.NASA CEA 600 code, Computer Program for Calculation of Complex Chemical Equilibrium Compositions and Applications. https://ntrs.nasa.gov/search.jsp?R=199 60044559 (1996)
2. Turns, An Introduction to Combustion: Concepts and Applications, Mcgraw-Hill Series in Mechanical Engineering (1999)

Chapter 5
Ramjets Fuels

Abstract In the past years, many fuels were tested for ramjet applications [1, 2] Most important features in the selection of fuels for ramjet applications are: 1. specific impulse (the amount of propellant required to accelerate the missile). 2. fast fuel/air kinetics to allow high combustion efficiency and short combustors. 3. density 4. storability 5. cost 6. storage safety. In fact, higher specific impulse allows higher thrust per unit weight of fuel, this saving the onboard propellant. When a long range is required, the higher Isp, the smaller fuel mass and consequently the lighter overall mass of the missile is required, yielding a greater acceleration for a given thrust. In air-to-air (small) missiles, the specific impulse per unit volume is critical for the shape design and lodging. Storable liquid fuels, e.g., liquid hydrocarbon (LHC), are attractive for their high density. In order to avoid instable and inefficient combustion, injectors may be able to allow vaporization and mixing of the liquid fuel with air in a very short time. The fast kinetics speed up the combustion improving its efficiency, i.e., increasing the amount of burned fuel that contributes to the thrust. The rate of reaction depends on the combustor inlet temperature and pressure.

5.1 Proprieties of Ramjet Fuels

The first fuel investigated for ramjet applications was JP-4 [1, 2]: this fuel consists of a mixture of kerosene (40–50%) and petrol (50–60%), characterized by a low flashpoint (23 °C) and a relatively high volatility (minimum boiling point of 61 °C): this allows its use even at low atmospheric temperatures. The lower volatility also involves higher risks of unintentional ignition. For this reason, JP-4 was replaced, over the years, by a fewer volatile fuel. One of these is the JP-5 [3], having a higher flash point (64 °C) and a chemical composition similar to JP-4. Improvements in pilot combustors and igniters allowed ordinary JP-5 jet engine fuel to be used in the US Navy Talos.

Widely used for ramjets propulsion was also the JP-7 [1]. It has excellent thermal and oxidative stability. The Jet Fuel Thermal Oxidation Test (JFTOT) stated that, at equal flow rates, the JP-7 releases deposits, on a test tube, at a temperature of 355

© The Author(s), under exclusive license to Springer Nature Switzerland AG 2021   35
A. Ingenito, *Subsonic Combustion Ramjet Design*,
SpringerBriefs in Applied Sciences and Technology,
https://doi.org/10.1007/978-3-030-66881-5_5

°C and for a process duration of about 5 h equivalent to those at only 260 °C and only 2.5 h of other JP fuels. Moreover, being almost totally free of sulfur (~0.001%) and aromatic compounds (~5%), the JP-7 provides highly efficient combustion and lower radiative heat transfer.

Since the '80 s, JP-8 was also used to replace JP-4 in certain applications. JP-8 is a kerosene-based fuel characterized by a flash point higher (53 °C) than JP-4 but also by a higher freezing point (54 °C). This makes it unsuitable for use in cold conditions. The last fuel of the JP family is the JP-10. It consists of a single molecule of exo-tetra-hydro-cyclo-pentadiene, having a low freezing point (79 °C) and a high volatility that allows a reliable ignition of the engine at high altitudes.

More recently, higher density fuels were valued as alternative fuels for ramjet applications. The RJ fuels family, starting from RJ1 (methylcyclopentadiene dimer), was in fact found to be a suitable substitute for JP-5, since it has a heat release 12% higher than JP-5. However, RJ1 polymerizes when stored in the missile fuel tanks, producing gum or varnish after a few months. The next fuel investigated was the RJ-4: this fuel does not polymerize gumming up ducts, and it produces less particulate residue than JP-5. Tests showed it increased maximum range by 20–30% and allowed a cruise altitude of 100,000 feet.

The next member of the RJ family that was investigated is RJ-4. This fuel is characterized by a low freezing point (46 °C) and by an energy per unit volume greater than the kerosene. One of the current fuels for ramjet propulsion is the RJ-5. This consists of dimers of hydrogenated perhydro-norbornadiene having high density (1.08 g/cm^3) and very low freezing point (29 °C) making its use at low temperatures [4, 5] problematic. Finally, RJ-6, composed of a mixture of RJ-5 (60%) and JP-10 (40%) is characterized by a volumetric heat of combustion lower than RJ-5 but, at the same time, by better chemical-physical properties at low temperatures. The freezing point of RJ-6 (54 °C) is far lower than that RJ-4 and RJ-5.

In Tables 5.1 and 5.2, some properties of these fuels are shown:

These two tables show that the RJ fuels are characterized by high density, higher fuel/air stoichiometric ratios, lower H/C ratios and therefore lower reaction heat.

Table 5.1 Density and reaction heat of some fuels

Fuel	Density (g/cm^3)	Heating value (MJ/kg)
JP-4	0.76	43,500
JP-5	0.81	43,025
JP-7	0.79	43,890
JP-8	0.81	43,140
JP-10	0.94	42,106
RJ-4	0.93	42,300
RJ-5	1.08	41,300
RJ-6	1.02	41,500

Table 5.2 Ratio H/C and stoichiometric ratio of some fuels

Fuel	Ratio	Stoich. ratio
JP-4	2.00	0.067558
JP-5	1.90	0.068290
JP-7	2.07	0.067194
JP-8	1.91	0.068589
JP-10	1.60	0.070762
RJ-4	1.67	0.070300
RJ-5	1.31	0.072638
RJ-6	1.42	0.072171

5.2 Fuel Performance Comparison

Assuming an ideal cycle of a ramjet engine that operates at a flight altitude of 15,000 m ($T_0 = 216.65$K), and with a maximum temperature of the cycle equal to $T_4 = 2500$K, by applying the previous equations for different fuels, specific thrust and fuel consumption as a function of the Mach number are shown in Figs. 5.1 and 5.2.

Table 5.3 shows a comparison of the characteristic velocity of the fuel, c*, the specific fuel consumption, TSFC, for different fuels and fuel fractions, ff.

Once fixed the maximum and minimum temperatures of the cycle, the specific thrust and the specific consumption of the ramjet increase by decreasing reaction heat of the fuel (see Figs. 5.1, 5.2, and Table 5.2). In fact, the smaller the amount of heat release, the greater the amount of fuel that must be burned to achieve a fixed

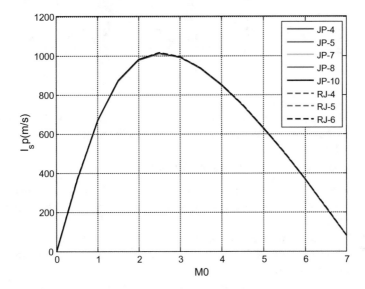

Fig. 5.1 I_a versus M for different fuels

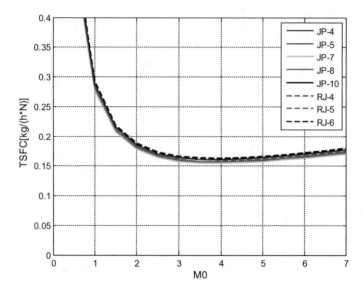

Fig. 5.2 TSFC versus Mach number for different fuels fuel

Table 5.3 C*, TSFC and mixing ratio of fuel for optimum Mach

Fuel	C* (m/s)	Ff	TSFC (kg/h/N)
JP-4	1010.44	0.046473	0.165576
JP-5	1011.29	0.046986	0.167261
JP-7	1009.75	0.046060	0.164216
JP-8	1011.08	0.046861	0.166850
JP-10	1013.01	0.048012	0.170624
RJ-4	1012.64	0.047792	0.169904
RJ-5	1014.57	0.048949	0.173685
RJ-6	1014.18	0.048713	0.172915

T_4. These differences, however, are almost negligible. Therefore, for a particular application, the choice of the fuel is driven mainly by its physical properties, as reported in Table 5.3.

Summarizing the fuels main features (Table 5.4):

- JP-4, JP-5, JP-7 and JP-8 have a boiling range (and thus a volatility) higher compared to other fuels such as JP-10 (whose boiling range is practically zero), RJ-4 and RJ-5;
- JP-4, JP-8 and JP-10 have freezing points lower than the RJ family fuels;
- the JP-4 is highly flammable; his flammability point (-23 °C) is much lower than that of all other fuels;
- fuels (except RJ-4) have roughly the same autoignition temperature

Table 5.4 Physical properties of fuels (at 1 atm)

Fuel	Freezing point (°C)	Flammable range (°C)	Boiling range (°C)	Autoignition point (°C)
JP-4	−62	−23–18	61–239	246
JP-5	−48	64–102	182–258	241
JP-7	−44	60–100	189–251	241
JP-8	−54	53–77	167–266	238
JP-10	−79	53	182	245
RJ-4	−46	71	207–221	329
RJ-5	−29	104	260–285	234
RJ-6	−54	61	182–285	232

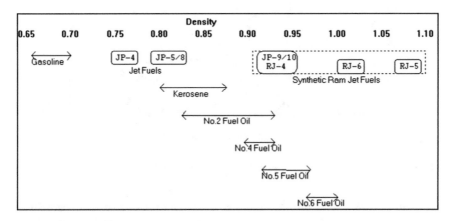

Fig. 5.3 Typical densities for aviation fuels [5]

- Even if we consider the speed of the flame front, the various fuels have different characteristics.

  Representative densities are shown in Fig. 5.3.

5.3 Ignition Delay Time for Ramjet Fuels

The combustion of a mixture does not occur instantly, but after a measurable time: the time elapsed from the spark to the moment the flame front appears is the ignition delay time. The ignition delay time is a critical parameter, defined in a rather practical way, as the time required for the temperature to rise more than 400 K after the fuel injection. Numerous studies have been done to extrapolate correlation laws of ignition delay as function of the initial temperature, mixture ratio and pressure.

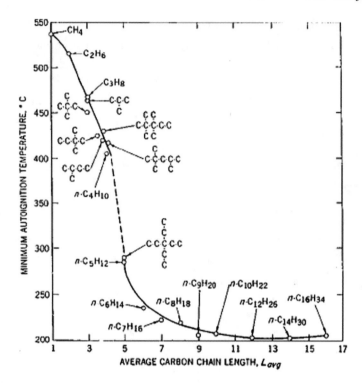

Fig. 5.4 Minimum auto-ignition temperature of hydrocarbons [6]

The figure below shows the auto-ignition temperatures of some hydrocarbons at atmospheric pressure. It is clear that the ignition delay time increases by increasing the hydrocarbon chain length. In fact, before ignition, in order to react, fuel must decompose and recombine with the oxidizer. Shorter is the chain length, shorter is the decomposition time (Figs. 5.4 and 5.5).

The autoignition temperature is the lowest temperature for the ignition, without any external energy source. This temperature decreases by increasing the chain length and the pressure.

In [7–11], the effects of inlet-air pressure and velocity on the Jet A ignition delay time for different temperatures are shown. At a constant velocity of 21.4 m/s, increasing the inlet-air pressure from 2 to 20 bar decreased the autoignition temperature limit from ~700 to 555K. At a constant inlet air pressure of 4 bar, increasing the reference velocity from 12.2 to 30.5 m/s increased the spontaneous ignition temperature limit from approximately 575 to 800K(Figs. 5.6 and 5.7).

During the ignition lag, the pressure does not increase greatly. At the end of the ignition lag, pressure increases rapidly ($\sim 10^4$ bar/s) and then decreases when the entire mixture is burnt (see Fig. 5.7) since the pressure losses are higher than the pressure increase due to combustion.

Conventional empirical laws predicting ignition delay are expressed as:

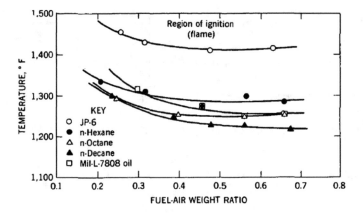

Fig. 5.5 Auto-ignition temperature as function of fuel concentration [7]

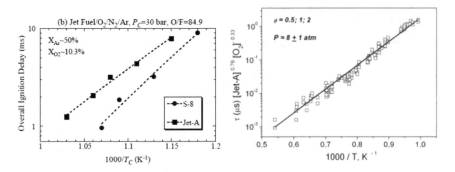

Fig. 5.6 Ignition delay time as function of pressure and temperature for two fuels [11]

$$\tau = A \cdot P^{-n} \cdot e^{E/RT} \tag{5.1}$$

where T is the temperature in Kelvin, E is the activation Energy (in kcal/mol), P is the pressure in atmosphere, A and n are empirical constants, τ is the ignition delay time (in ms).

The ignition delay time for JP-10 is given by Kundo as:

$$\tau_{acc} = 7.63 \cdot 10^{-16} e^{46834/RT} [JP-10]^{0.4} [O_2]^{-1.2} \tag{5.2}$$

In [10], empirical constants for JET A and JP4 are reported (see Table 5.5). These constants have been validated in the range of $650 - 900$ K, $10 - 30$ atm, and $\varphi = 0.3 - 1$.

Figure 5.8 shows that increasing pressure from 1 to 10 atm decreases the ignition delay of JP4 by 90 to 99% and of JET A by 90%. Increasing φ from 0.15 to 0.9

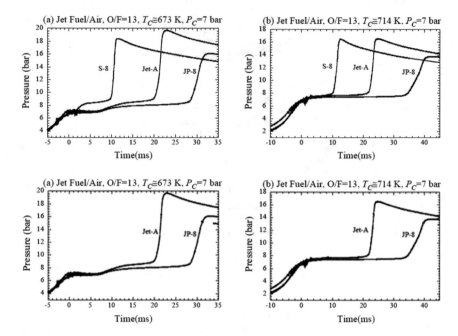

Fig. 5.7 Pressure history [12]

Table 5.5 Coefficients for ignition delay time

Fuel	n	A	E kcal/mol
Jet-A	2	1.60×10^{-8}	37.78
JP-4	1	6.89×10^{-9}	35.09

decreases the ignition delay by nearly 50%. The increase of temperature instead decreases the delay by almost 300% [12].

Experimental results in [13] show that ignition delay of stoichiometric JP-10/air mixtures at 1750 K and 20 atm [4] are about 100ms, i.e., longer than that of the JP-8 at the same conditions (0.1 ms).

5.4   Flame Speed for Ramjet Fuels

The laminar flame speed is the speed at which the flame propagates through a mixture of unburned reactants [11]. Mallard and Le Chatelier [11] stated that the *laminar* flame speed depends on the thermal diffusivity α, the reaction rate $\dot{\omega}$ and the temperature through the flame zone:

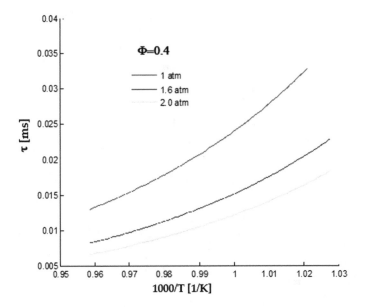

Fig. 5.8 Ignition delay time of JET A versus $1000/T$

$$s_L^o = \sqrt{\alpha\dot\omega\left(\frac{T_b - T_i}{T_i - T_u}\right)} \qquad (5.3)$$

where u is for unburned, b is for burned and i is for ignition temperature.

Nevertheless, the combustion regime in a RJ combustor is certainly turbulent (in a RJ combustor at M ~ 0.3, the Reynolds number is of order 10^5), the structure of the reaction zone at high Reynolds number is likely based on laminar flamelets or associated to them.

In [14, 15], flame speeds of JP8 and JP10 were measured at 600–682 K, assuming stoichiometric conditions, $p = 1$ atm.

The flame speed of JP8 is higher than JP10 (see Fig. 5.9) and both increase with the mixture initial temperature. The flame speed also has a maximum for near stoichiometric conditions. Figure 34 shows flame speed of JP-7 and JP-8 for different fuel fractions, calculated assuming an initial temperature of 403 K and a pressure of 1 atm. JP7 achieves higher laminar flame speeds than JP8 and consequently than JP10 (Fig. 5.10).

Concluding:

- for applications requiring maximizing the specific thrust, RJ fuels are advisable, but they have rather high freezing points that make them unusable at low temperatures.
- for applications that require to maximize combustion efficiency, JP-7 is recommended since it burns stably and has burning rate exceeding that of the other JP fuels.

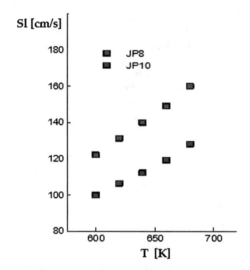

Fig. 5.9 Laminar flame versus initial temperature

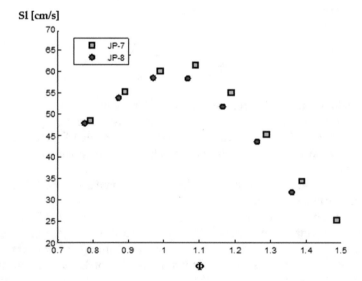

Fig. 5.10 Sl versus φ

- for applications that require a fuel with good performance even at low temperatures, JP-4 is recommended because of its high volatility; this, however, encom-
  passes the risk of accidental ignition and therefore is not recommended on board
  ships.
- at high altitude JP-10 does not have enough volatility to ensure reliable ignition.

Therefore, depending on the application, the optimum choice of fuel is the result of the compromise between chemical–physical characteristics and performance.

References

1. C.R. Martel, MILITARY JET FUELS, 1944 – 1987, AFWAL-TR-87-2062 (1987)
2. Aviation Fuel Proprieties, CRC Report No. 530 (1983)
3. T. Edwards, Liquid fuels and propellants for aerospace propulsion: 1903-2003. J. Propul. Power **19**(6), 1089–1095 (2003)
4. J.L. Ross, A fuel data standardization study for JP-4, JP-5, JP-7 and RJ-5 Combusted in Air, AFAPL-TR-74-22, Ohio (1974), pp. 1–4
5. P.G. Hill, C.R. Peterson, Afterburner and ramjet combustor, Mechanics and Thermodynamics of propulsion (1992), pp. 256–262
6. M.G. Zabetakis, Flammability characteristics of combustible gases and vapours, U.S. Department of Mines, Bulletin 627 (1965)
7. R.D. Ingebo, C.T. Norgren, Spontaneous-ignition temperature limits ofjet a fuel in research-combustor segme, (NASA-TM-X-3146)
8. A. Mestre, F. Ducourneau, Recent Studies on the Spontaneous Ignition of Rich Air-Kerosene Mixtures. ONERA No. 1209, Office National d'Etudes et de Recherches Aerospatiales (France) (1973)
9. R.D. Ingebo, A.J. Doskocil, C.T. Norgren, High-Pressure performance of combustor segments utilizing pressure-atomizing fuel nozzles and air swirlers for primary-zone mixing. NASA TN D-6491 (1971)
10. H.C. Barnett, R.R. Hibbard, eds.: Basic Considerations in the Combustion of Hydrocarbon Fuels With Air. NACA TR-1300 (1957), pp. 106–107
11. NASA Contractor Report 159886 Autoignition Characteristics of Aircraft. Type Fuels, By Louis J. Spadaccini, John A.TeVelde Prepared by United Technologies Research Center East Hartford, CT 06108 For National Aeronautics and Space Administration, Lewis Research Center Cleveland, Ohio June 1980
12. G. Francesco, Analisi comparativa di combustibili per applicazioni Ramjet. Master degree thesis, Italy (2015)
13. E. Mallard, H. L. Le Chatelier, Ann. des Mines (8) 4, 274 (1883)
14. NASA Contractor Report 175064 Spontaneous Ignition Delay Characteristics of Hydrocarbon Fuel/Air Mixtures, Arhur Lefebvre,W. Freeman, and L.Cowell Purdue University West Lafayette, Indiana February 1986
15. E.F. Kian, Burning speeds, flame kernel formation and flame structure of bio-jet and JP-8 fuels at high temperatures and pressures, Ph.D. thesis, Northeastern University, Boston, Massachusetts (2010), pp. 203–233

Chapter 6
Flameholder Design Guidelines

Abstract Flameholders are a key component in ramjet combustor. They are designed to increase airflow turbulence and therefore turbulent recirculation ensuring efficient combustion within a short distance, anchoring and maintaining the flame stable. Flameholders create an eddying region that provides enough residence time for fuel and air to mix and burn, preventing the flame from being blown out. Their geometry should allow the rising of stable eddies preventing, at the same time, excessive drag increase; a proper flameholder configuration is mandatory to maximize the ramjet performance.

6.1 Flameholder Geometries

Different sizes and geometries have been analyzed over the years: cylindrical, discs, "V" and "H" shaped annuli… A typical configuration is shown in Fig. 6.1.

In a real combustor, pockets of unburned mixtures are transported downstream, beyond the turbulent flame front. Even if the apparent turbulent flame front can "sweep" the entire cross-section of the combustor and air/fuel flow, additional length is required to complete the combustion process (Figs. 6.2 and 6.3).

If a similar burner to that above but having a single gutter is considered, since the length required for the flame propagation increases, the length necessary to burn pockets of unburned mixture decreases, with a corresponding loss of combustion efficiency.

This suggests that, for a burner of limited length, increasing the number of gutters increases the efficiency of combustion. However, experimental studies in [1] showed that an increase of the V gutter number over three is not advantageous due to the increase of pressure losses. Further, the increase of the flameholder cross-section may force large velocity along the edges of the flameholder that cause the flame to break away (flame blow-out). In [2], the optimum cross-section is indicated as 1/3 for a conical flameholder and 1/2 for a ring-shaped flameholder.

Detailed research on flame stabilization was also carried by De Zubay [3] over a wide range of mixture compositions, pressures and flow velocities (from 0.2 to 1atm

© The Author(s), under exclusive license to Springer Nature Switzerland AG 2021 47
A. Ingenito, *Subsonic Combustion Ramjet Design*,
SpringerBriefs in Applied Sciences and Technology,
https://doi.org/10.1007/978-3-030-66881-5_6

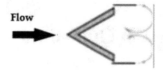

Fig. 6.1 Flameholder V-gutter

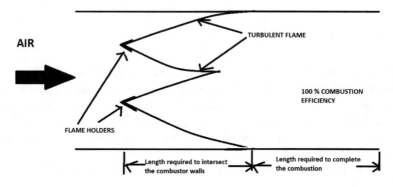

Fig. 6.2 Two 2D flameholder V-gutter

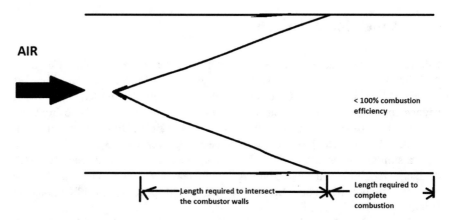

Fig. 6.3 Single flameholder V-gutter

and from 12 to 170 m/sec). De Zubay found that the blowout limit is a function of the stabilization parameter.

$$\frac{F}{A} = f\left[\frac{w}{p^{0.35}D^{0.85}}\left(\frac{T_0}{T}\right)^{1.5}\right] \tag{6.1}$$

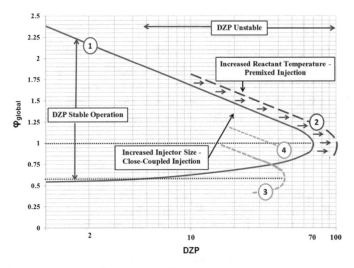

Fig. 6.4 Flameholder stability limits

where T_0 is the temperature at which experiments have been done, in K, T is the temperature of the flow in K, p is the pressure of the flow in atm or kg/cm^2, w is the velocity of the flow in m/sec, f is the De Zubay function, approximated by the curve in Figure, D is the hydraulic diameter of the flameholder in cm: $D = 4S/P$, S is the area of the flameholder in cm^2, P is the wetted perimeter in cm.

For example, assuming the stoichiometric fuel–air ratio for pentane/air, i.e., 0.0167, the stabilization parameter according to Fig. 6.4 is 70.

From this equation, it is possible to calculate the velocity, the pressure, and the temperature of the flow or the hydraulic diameter of the flameholder, at blow out condition, if the remaining parameters are known.

Assuming an inlet velocity $V = 100$ m/sec; the ratio between static and total temperature $T/T_0 = 1$; $p = 1$ kg/cm^2, the smallest hydraulic diameter of a flameholder capable of maintaining the flame anchored is equal to $D_{in} = 9.4$ mm.

Heating of the flameholder expands the stabilization limits at a given velocity and a given composition of the mixture. Cooling of the flameholder, instead, shrinks the stabilization limits. Understanding of the stability of the flame is also due to Longwell et al. [4]. Ballal and Lefebvre [5] proposed a correlation between the equivalence ratio blow out and the temperature, T, pressure, p, velocity, U, turbulence intensity, T_u, flameholder size, D, and flameholder blockage ratio B:

$$\phi_{LBO} = \left[\frac{2.25(1 + 0.4U(1 + 0.1T_u))}{p^{0.25}\left(e^{T/150}\right)D(1 - B)} \right]^{0.16} \tag{6.2}$$

They concluded that a primary influence on the weak extinction limits is the inlet temperature.

In 1988, Stanley and Lefebvre and Kim [5] conducted a research on "irregular" shaped flameholders concluding that there is no appreciable impact on the extinction, and that the flameholder size is the most important parameter in aerodynamic blockage; the shape of the flameholder instead plays "a very minor role." They recognized in the U/D ratio the Damköhler timescale:

$$Da_{ig} = \frac{D}{U\tau_{ig}} \tag{6.3}$$

If the ignition delay time is longer than the residence time (Da < 1), the reactants will not be able to ignite, causing blowout. Therefore, a correlation for the lean blow off limit accounting for the Damköhler number was defined:

$$\phi = \frac{0.1950(U/D)^{0.1490}e^{\frac{361.4101}{T}}}{p^{0.2199}} \tag{6.4}$$

6.2   Theoretical Analysis of V-type Flameholder Geometry

Flameholders have proved their ability to accelerate the flame anchoring improving the combustion efficiency, though, this is at expenses of the pressure losses. Figure 6.5 shows a simplified scheme of the flow conditions next the flameholders ensuring stable flame anchoring.

In [6], a quasi-one-dimensional analysis has been suggested to predict the performance of different flameholders geometries (see Fig. 6.5). Different configurations are distinguished by two different parameters: the flameholder open area ratio A_f/A_1 and the area contraction coefficient K depending on the flameholder shape.

Fig. 6.5 Flameholder open area A_f [6]

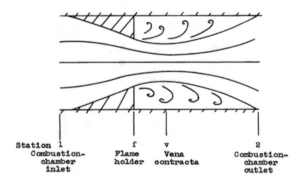

Station 1
Combustion-
chamber
inlet

f
Flame
holder

v
Vena
contracta

2
Combustion-
chamber
outlet

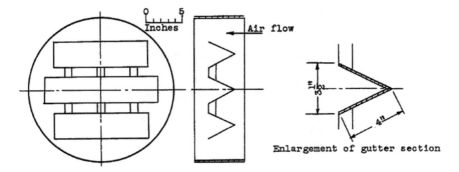

Fig. 6.6 Three horizontal V-gutter flameholders [7]

A_f is the flameholder open area, i.e., the difference between combustion chamber area A_1 and the cross-sectional flameholder area, namely the section through which the air stream flows, see Fig. 6.5. In this book, two different flameholder geometries have been accounted [7]: the three horizontal V-gutter and the two annular-V-gutter flameholders; for each of them different width and size have been analyzed.

For the first configuration (see Fig. 6.6), i.e., a three horizontal V-gutter flameholder, the contraction coefficient K has been experimentally determined in [7], and it is $K = 0.80$; In that each V-gutter has a width of 8.9 cm and a length of 33 cm except the central one, which is 39.6 cm long, therefore the flameholder cross-sectional area is 989.67 cm^2, and the combustion chamber cross-sectional area is 2025.80 cm^2. Consequently, $A_f/A_1 = 0.51$.

The other flameholders considered, shown in Fig. 6.7, are three annular-V-gutter geometries, different for width and number of rings. For this design, the contraction coefficient K, experimentally found in [7], is $K = 0.95$.

The vena contracta area is the product of the flameholder open area A_f and the area contraction coefficient K, which depends on the shape of the flameholder. Flameholders provide a sudden enlargement of the flow area driving recirculation of the airstream.

$$K\frac{A_f}{A_1} = \frac{M_1}{M_v}\left(\frac{1 + \frac{\gamma-1}{2}M_v^2}{1 + \frac{\gamma-1}{2}M_1^2}\right)^{\frac{\gamma+1}{2(\gamma-1)}} \tag{6.5}$$

Heat is gradually added as the gas stream flows through the combustion chamber from station 1 to station 2 (see Fig. 6.8). Due to the heat addition, changes in total pressure and velocity occur. The variation of combustion chamber outlet Mach number M_2 with the combustion chamber inlet Mach number M_1 due to combustion heat assumed released at station 1, is given by [8]:

$$\left(\frac{R_2}{R_1}\right)\left(\frac{T_2}{T_1}\right) = \left(\frac{M_2}{M_1}\right)^2\left(\frac{\gamma_2}{\gamma_1}\right)\left(\frac{1 + \gamma_1 M_1^2}{1 + \gamma_2 M_2^2}\right)^2\left(\frac{1 + \frac{\gamma_2-1}{2}M_2^2}{1 + \frac{\gamma_1-1}{2}M_1^2}\right) \tag{6.6}$$

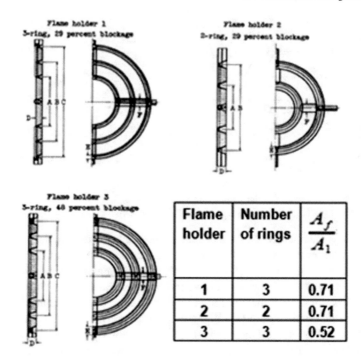

Fig. 6.7 Three annular-V-gutter flameholders [7]

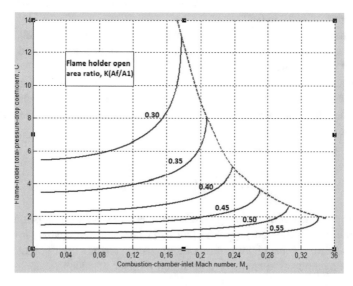

Fig. 6.8 Total pressure losses versus M_1 for different $K\left(A_f/A_1\right)$

R_2, R_1 and γ_1, γ_2 are the specific gas constant and heat capacity ratio for the stations 1 and 2.

Introducing a static pressure drop coefficient defined as $(p_1 - p_2)/q_1$ and using the conservation of momentum, following equation is derived:

$$\frac{p_1 - p_2}{q_1} = 2\left(\frac{1 + \gamma_1 M_1^2}{1 + \gamma_2 M_2^2}\right)^2 \left(\frac{M_2}{M_1}\right)^2 \left(\frac{\gamma_2}{\gamma_1}\right) - 2 \tag{6.7}$$

where p_1 and p_2 are the inlet and outlet static pressure and q_1 is the inlet dynamic pressure.

Applying the equations of conservation of mass, momentum and energy, the following equations, which relate the total pressure ratio to the effective open area of the flameholder and the combustion chamber inlet Mach number, can be obtained.

$$C = \left(1 - \frac{P_2}{P_1}\right) \frac{\left(1 + \frac{\gamma-1}{2} M_1^2\right)^{\frac{\gamma}{\gamma-1}}}{\frac{1}{2}\gamma M_1^2} \tag{6.8}$$

$$M_2^2 = \frac{P_1}{P_2}\left(\frac{1 + \frac{\gamma-1}{2} M_2^2}{1 + \frac{\gamma-1}{2} M_v^2}\right)^{\frac{\gamma}{\gamma-1}} \left(K\frac{A_f}{A_1}M_v^2 + \frac{1}{\gamma}\right) - \frac{1}{\gamma} \tag{6.9}$$

The total pressure drop coefficient C has been defined by:

$$C = \frac{P_2 - P_1}{q_1} \tag{6.10}$$

where P_1 and P_2 are the inlet and outlet section total pressure and q_1 is the inlet dynamic pressure.

Solving this system of equations, the variation of $(p_1 - p_2)/q_1$ with M_1 and $(R_2/R_1)(T_2/T_1)$ for $\gamma_1 = 1.4$ and $\gamma_2 = 1.3$ can be obtained.

For low open area ratio $K(A_f/A_1)$, Figure shows a significant increase of the coefficient C with the inlet Mach number M_1, reaching the thermal chocking (red line). Instead for $K(A_f/A_1) > 0.45$ the coefficient C results almost constant with the increase of M_1; pressure losses are lower and the sonic flow conditions (choking) are reached at higher values of M_1.

In order to reduce the pressure losses due to flameholder drag and combustion heat addition, it is necessary to have high values of flameholder open area ratio; this condition may be satisfied with a imposed maximum blockage ratio of 0.35.

Figure 6.9 shows that for $(R_2/R_1)(T_2/T_1) < 2$ the static pressure drop coefficient $(p_1 - p_2)/q_1$ does not vary much with the inlet Mach number M_1, while for $(R_2/R_1)(T_2/T_1) > 2$ the coefficient $(p_1 - p_2)/q_1$ increases with M_1. For $(R_2/R_1)(T_2/T_1) = 3$ and $M_1 = 0.28$ chocking takes place. In order to retard chocking a lower Mach number at the combustor inlet should be assumed. For

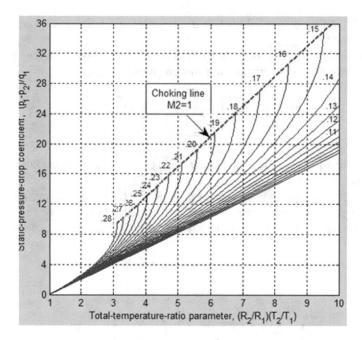

Fig. 6.9 Static pressure losses versus $(R_2/R_1)(T_2/T_1)$ for different M_1

example, when $(R_2/R_1)(T_2/T_1) = 6$, Mach numbers lower than 0.19 are mandatory to avoid thermal chocking. Further, pressure losses increase with the inlet Mach
number and with the heat added by the fuel.

Therefore, it is critical to define a right balance between high total temperature
ratio, which means high performance, and the resulting high-pressure losses, which
reduce this performance. It is also interesting to see how increasing the inlet Mach
number, choking conditions, represented in the figure with the red line, are reached
earlier.

6.3 Flame Stability

Due to the flameholders, continuous combustion occurs in the recirculation area
behind them. The cold, turbulent flow of air and fuel, often partially premixed and
flowing around the flameholder, comes into contact with the burning gas in the
recirculation region, and by means of turbulent mixing heat transfer obtains the
heat necessary for its ignition. If the velocity of the cold mixture is high then the
flame speed or if the mixture temperature is low, the quantity of heat that the fresh
mixture receives will be insufficient for heating it to the ignition temperature T_{ig}, and
the mixture will not ignite: the flame will blow away from the flameholder (flame
blowout, or blowoff).

During tests, a flameholder with a known relative cross-section is installed in a combustion chamber and the velocity at the chamber inlet is increased until combustion takes place. At a given relative cross-section, the flameholder that has the highest blow off velocity is the best.

The geometry of a flameholder exerts a complex influence upon the blow out velocity. The blow out velocity increases with an increase of the flameholder's perimeter so long as the transverse dimensions of its elements do not become less than a certain given value. Therefore, complicated flameholders, constructed from radial or concentric gutters or of rods, retain the flame better than a conical holder if the loads applied to the cross-section are equal. The blow out velocity increases with an increase of the rods cross-section; that, however, also increases the blockage.

Flameholder width influences the flame stability. The stability limits have been correlated with flameholder, or step, height, by introducing the Damkohler number at blowout for premixed conditions: this is the ratio between fluid dynamic, or residence, time to chemical kinetics time. ϕ is the [assumed] premixed fuel–air equivalence ratio, H is the step height, U is the axial velocity of gas and τ_{PREMIX} is the characteristic chemical time (Fig. 6.10).

The upper branch of the curve represents the rich stability limit, while the lower represents the lean stability limit.

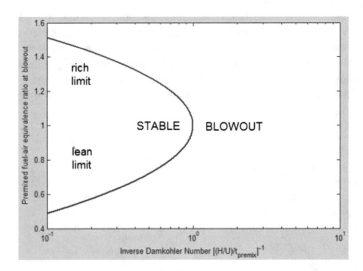

Fig. 6.10 ϕ versus inverse of Damkohler number

6.4   Investigation of Flameholder Blockage Area Ratio

In [6, 7], an experimental investigation of the effect of mainstream velocity, blockage ratio, and flameholder shape on the lean limit of flammability and efficiency was performed. These tests investigated different blockage ratios, from 0.25 to 0.5, and different shapes of bluff body (cone, flat plate, cylinder, and sphere). The tests showed that:

1. increasing the flow velocity has an adverse effect on the flammability lean limit
2. increasing the blockage ratio widens the range of stability for cylindrical and spherical flameholders, but has an opposite effect in case of cone and plate
3. the shape of the bluff body affects the recirculation region and consequently the lean limit.

   A summary of the Literature Review is reported in Table 6.1.
   In the next section, different flameholder geometries are investigated.

6.5   Some Examples: Multistage V-gutter Flameholders

This study was carried out between 1947 and 1949 at what was then the NACA Lewis Laboratory [7], but it is still instructive, as it shows how the problem of maintaining the flame stably anchored while ensuring combustion efficiency was attacked. In these experiments, it was possible to obtain data only for a single operating condition, due to the temperature limitations of Inconel alloys. Given the very short life of the tested pieces (about 4 min), flameholders were coated with molybdenum silicide. This extended the life of the flameholders during the tests up to 47 min.

   Initial data (pressure, temperature and speed relate to the condition of the air at the burner entrance):

- $Fuel/air\ ratio\ \phi = 0.05$
- $Residence\ time\ t = 0.01\mathrm{s}$
- $Static\ pressure\ P < 203182.72\,\mathrm{Pa}$
- $Velocity\ U = 60.96\,\mathrm{m/s}$
- $Combustor\ Length\ L = 0.61\,\mathrm{m}$
- $Temperature\ T = 366.48\,\mathrm{K}$
- $0.18\,\mathrm{m} < transversal\ section < 0.2\,\mathrm{m}$
- $Fuels: iso-pentane.$

   The configurations analyzed are reported in the following.

Table 6.1 Literature review

Reference	Experimental data	Results
Longwell et al. [9]	• BR = 0,02 to 0.23, cylinder, cone and V-gutter • U = 69 to 274 m/s T = 339 to 533 K • P = 0.1 to 3.2 atm • Naphtha/air	• Increasing U, decreasing T and/or d and streamlining trailing edge of baffle decrease stability range • Pressure unimportant
Williams et al. [10]	• BR = 0.0005 to 0.17 rod, • V-gutter and flat plate • U = 6 to 107 m/s • Tu = 0.4 to 80% • T = 300 to 340 K • Natural gas/air	• Increasing U or Tu decreases BR and/or cooling of the flameholder decreases the stability range • -Baffle shape: unimportant
Wright [11]	• BR = 0.03 to 0.26 flat plate • U = 37 to 185 m/s • P = 1 atm • gasoline/air	• The flame speed and geometry depend on blockage • The blowout velocity was given by: $U = \dfrac{BR}{C1+C2 \cdot BR}\dfrac{h}{r}$ which predicted that max blowout speed would be C1/C2: for plate 0.35 and for cylinder 0.56 • The chemical time does not depend on flameholder
Lefebvre [12]	• BR = 0.11 to 0.44 cones • U = 41 to 134 m/s • T = 293 to 774 K • butane/air	• Increasing U, decreasing T and/or increasing BR increases $\dfrac{\dot{M}_E}{M}$ • For certain BR, pipe and baffle diameter unimportant • $\dfrac{\dot{M}_E}{M} =$ $0.65\left(\dfrac{U}{t^{0.75}}\right)\left[\dfrac{BR}{(1-BR)^{0.5}}\right]^{1.5}$
Ballal and Lefebvre [13]	• BR = 0.04 to 0.34 cones, • U = 10 to 100 m/s • Tu = up to 15% • T = 300 to 575 K • P = 0.2 to 0.9 bar • propane/air	• Increasing U and/or Tu increases ϕ, while increasing BR and/or T has an adverse effect • Pressure: unimportant • $\phi =$ $\left\{\dfrac{2.25[1+0.4U(1+0.1Tu)]}{P0Te^{T/150}d(1-BR)}\right\}^{0.16}$

(continued)

Table 6.1 (continued)

Reference	Experimental data	Results
Baxter and Lefebvre [14]	• BR = 0.125 to 0.32 v-gutter • Mn = 0.18 to 0.26 • T = 65' to 850 K • W = 25.4 to 65.1 mm • ϑ = 45, 60, 90° aviation kerosene, • JP5/air	• Increasing U decreases the stability range • Stability increases due to increasing gutter width • The shape of bluff body affect its stability characteristics • Any increase of T widens the range of the stability
Walburn [15]	• BR = 0.083 to 0.167 cylinder • U = 30 to 100 m/s • Propane/air	• Heterogeneity of the reaction zone and a progressive increase in reaction efficiency downstream of the stabilizer
Pan et al. [16]	• BR = 0.13 to 0.25 solid cones • ϑ = 30 to 90 deg, • methane/air	• Increasing BR slightly decreases L but increases the shear stress and TKE • Increasing Tu shortens L to its cold flow value

6.5.1 Two Parallel Lines of Gutters

The two parallel lines of gutters flameholder geometry are shown in Fig. 6.11. The two parallel lines of 11 gutters are spaced with their centers 12.7 cm apart. These gutters were 10 cm long, 2.54 cm from tip to base, 1.9 cm wide and 3.8 cm between tips.

Combustion limits due to blowout are given in Fig. 6.12. Combustion was found stable even at inlet air velocity of 99.06 m/s and a fuel air ratio of 0.07, and at 60.96 m/s for $f = 0.018$. However, combustion efficiency was below 30% at the standard operating conditions.

Despite stability was adequate, the combustion efficiency was unsatisfactory. In order to analyze whether any of the dimensions of this flameholder significantly affect

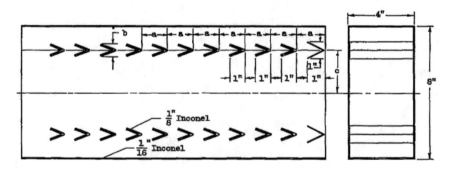

Fig. 6.11 Two parallel lines of gutters [7]

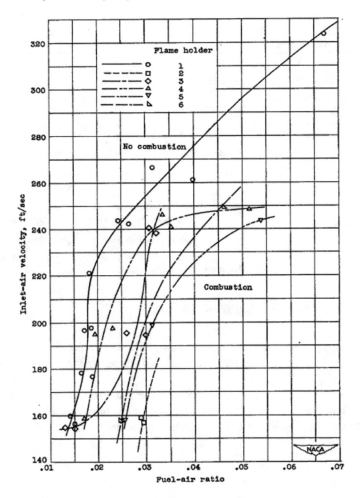

Fig. 6.12 Combustion limits for Two parallel lines of gutters [7]

the combustion efficiency, different configurations have been analyzed. Difference between these configurations is reported in Table 6.2.

Increasing the number of gutters from 11 to 15 decreases the flammability range. In fact, as the number of gutters increases also the mixture speed increases and therefore, the flammability range decreases as well. Increasing the distance and the size of each gutter does not significantly increase the combustion efficiency that staid below 30%.

Table 6.2 Flameholder geometry

Flameholder	Dimensions cm	Total gutters
1	a = 3.8; b = 1.9; c = 6.35	22
2	a = 6.35; b = 1.9; c = 6.35	14
3	a = 2.7; b = 1.9; c = 6.35	30
4	a = 3.8; b = 1.9; c = 5	22
5	a = 3.8; b = 2.54; c = 6.35	22
6	Same as flameholder 1 except for addition of third row of gutters in center	33

Fig. 6.13 Three parallel lines of gutters [7]

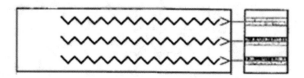

6.5.2 Three Parallel Lines of Gutters

A third row of gutters was inserted among those listed in the previous configuration (flameholder 6 in Fig. 6.12. In this test, vaporized iso-pentane and liquid AN-F-48b were used (Fig. 6.13).

At a combustor pressure of $186250.83 Pa$, the combustion efficiency increased and was found to be of 50% for both fuels. However, the flammability range was lower than the flameholder with two parallel lines of gutters.

6.5.3 Staggered Gutters

This configuration contains 11 gutters 0.6cm alternately staggered from the row centerline. Only the AN-F-48b liquid is used as fuel. This configuration showed higher efficiency than any other two row configurations (Fig. 6.14).

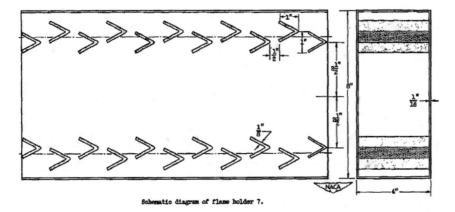

Schematic diagram of flame holder 7.

Fig. 6.14 Staggered gutters [7]

6.5.4 Three Rows of Staggered Gutters

This configuration combines features of the two configurations of flameholder shown above, i.e., flameholder 6 (using 11 gutters), and staggered gutters (Fig. 6.15).

With this configuration, combustion efficiency of 65% using vaporized isopentane and at inlet air temperature = 410.93K was obtained.

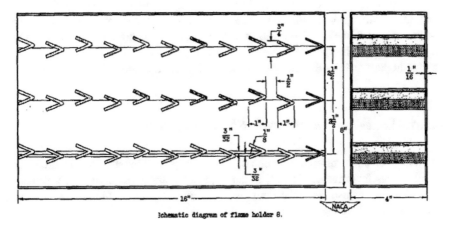

Schematic diagram of flame holder 8.

Fig. 6.15 Three parallel lines of gutters [7]

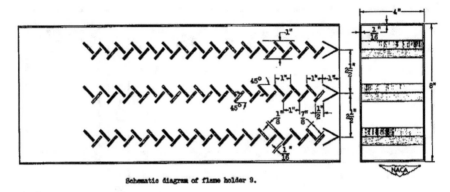

Fig. 6.16   Three parallel lines of flat plates [7]

6.5.5   Other Configurations

Next, flameholders in which the gutters were replaced with 30 staggered flat plates provided with holes to allow a continuous flame propagation between the regions of low speed were investigated (Fig. 6.16).

After 4 min of activity, these tests showed a combustion efficiency of 57% at air inlet temperature $388.71 K$ and pressure $189637.21 Pa$ with liquid isopentane. Such flameholders were subjected to rapid melting.

Flameholder 10 is composed of three corrugated strips with holes that allow a continuous flame propagation between the regions of low speed and was tested with the use of isopentane as a vaporized fuel. The condition of choking that is created in the nozzle limits the speed of the air to $45.72\,\mathrm{m/s}$. In this condition and with an air inlet temperature $416.48 K$, the combustion efficiency reached 80%.

6.5.6   Rake- and Gutter-Type Flameholders

The effect of other different flameholder geometries on combustion efficiency and stability was investigated. In [17] gutter–type, two stages (involving an upstream gutter and a downstream rake configuration) and rake-type configurations were experimentally investigated.

Conditions of experiments (temperature and Mach number relate to the condition of the air at the entrance to the burner) are:

- $3048m < fightaltitude < 10668m$;
- $273.15K < Temperature < 338.71K$;
- $0.49 < Mach < 1.46$;
- $Combustor\,length$ 2.93 m$(diameter$ 0.46 m$)$;
- $0.006s < ResidenceTime < 0.017s$;
- $0.04 < ratio\,fuel/air < 0.07$;
- Pressure: chosen based on the simulated flight altitude

Using rake flameholders and a mixture of **25%** propylene oxide and **75%** kerosene rather than kerosene, with a fuel / air ratio of 0.07, combustion efficiency increased from **59%** to **85%**.

The highest performance (**85%** efficiency) was obtained by the rake type (Fig. 6.17) flameholder, having 3 gutters inclined at 45° which connect the 3 rakes to the central ring (Fig. 6.18).

The best performance was found using kerosene and a mixture of **25%** propylene oxide and **75%** kerosene at the previous operative conditions, with $fuel/air$ ratio $= 0.07$. Combustion efficiency increased with inlet air temperature.

Figure 55 shows the combustion efficiency as function of equivalence ratio and fuel. This figure shows that in the temperature range examined, the best efficiency was **77%** for air speed 64.92 m/s and $f = 0.04$.

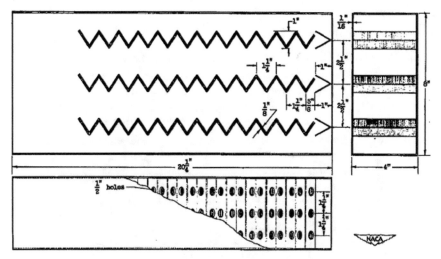

Schematic diagram of flame holder 10.

Fig. 6.17 Three parallel lines of corrugated strips with holes [7]

Fig. 6.18 Flameholder rake-type [17]

6.5.7   Grid Type Flameholder

Lastly, four different configurations of grid type flameholders were investigated: a 30° V-shaped standard type, a 3/4 of scale (Fig. 6.19b), a double scale, and a spaced 1.4 scale (Figs. 6.19d and 6.20).

Experimental operative conditions were:

- $0.042 < f < 0.051$;
- $Residence\,time = 0.002\,$s;
- $Mach\,numbers\,up\,to\,1.84$;
- $Combustor\,length = 1.52\,$m;

The fuel was the 62-octane AN-F-22 preheated to $366.5\,K$.

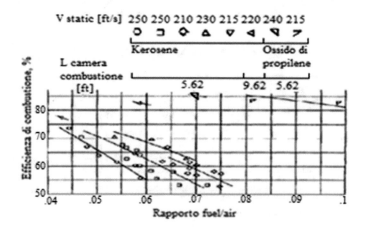

Fig. 6.19 Rake type flameholder combustion efficiency [17]

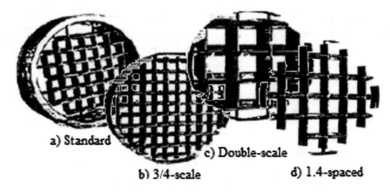

a) Standard

c) Double-scale

b) 3/4-scale   d) 1.4-spaced

Fig. 6.20 Grid type flameholder configurations [7]

Combustion efficiency was 50% with the standard configuration at an air inlet velocity and temperature respectively of $61.26\,m/s$ and $266.48\,K$, and for $f = 0.038$. Efficiency climbed to 75% with lower air speed (47.24 m/s) (Fig. 6.21).

The highest efficiency (78%) was obtained with a double scale (c) configuration and a $fuel/air\,ratio$ of 0.057. The air velocity could be raised up to 45.72 m/s. In fact, the efficiency decreased rapidly with increasing inlet air speed. Lower efficiency was found for the configuration d.

Concluding, experimental work on different configurations showed that the best performance is obtained for a corrugated flameholder, with an efficiency of 80% (using pre-vaporized isopentane).

As for rake-type flameholders, the best efficiency of **85%** was found with a rake type flameholder having 3 gutters (counted alternately) and inclined at 45° (using kerosene and a mixture 25% propylene oxide **75%** kerosene). As for the grid type flameholder, above **38.1** m/s, the best efficiency (**75%**) was obtained with the standard grid flameholder ($f = 0.05$). Below **38.1 m/s**, the best efficiency (**78%**) was obtained with a double scale grid flameholder.

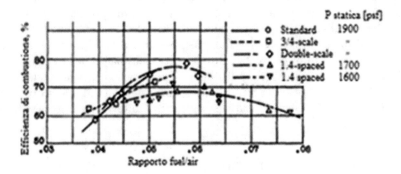

Fig. 6.21 Grid type flameholder combustion efficiency [7]

Table 6.3 Flameholder annular configurations details

Flameholder	Number of rings	Gutter width, cm	Blockage percentage
1	1	6.35	23
2	2	3.81	29
3	3	3.81	48
4	4	1.9	29

Table 6.4 Flameholder annular configurations dimensions

Flameholder	Type of gutters	Dimensions, cm
1	cambered	A = 40.44, D = 7.3, E = 6.5
2	V	A = 24.13, B = 47, D = 3.98, E = 3.81, F = 3.81
3	V	A = 26, B = 26, C = 57.5, D = 3.98 E = 3.81, F = 3.81
4	V	A = 25.7, B = 25.7, C = 57.5, D = 3.17, E = 1.9, F = 3.81

6.5.8 Annular Flameholders

In [18], combustion efficiency of annular flameholder was investigated. In general, the combustion efficiency increases with the number of annular gutters and the number of rings; however, the increase in the gutters number will also increase the blockage and therefore the total pressure losses. Tables 6.3 and 6.4 show the different configurations investigated (number of annular rings, the gutter widths, and the blockages of each of the flameholders) (Fig. 6.22).

6.5.9 Investigation of Flameholder V Angle and Width

Experimental results showed that the increase of the blockage ratio tends to increase combustion efficiency but at the same time increases the pressure loss. Further, increasing the width of the gutters significantly increases the combustion efficiency.

Figure 6.24 shows that combustion efficiency increases with the burner inlet velocity. Comparison of performance of the three rings flameholder configuration with $B = 48\%$ in Fig. 6.23 and the three rings flameholder configuration with $B = 29\%$ in Fig. 6.25, shows a higher blockage ratio allows a wide range of stability limits in terms of fuel air ratio and inlet air speeds. However, the price was higher pressure losses.

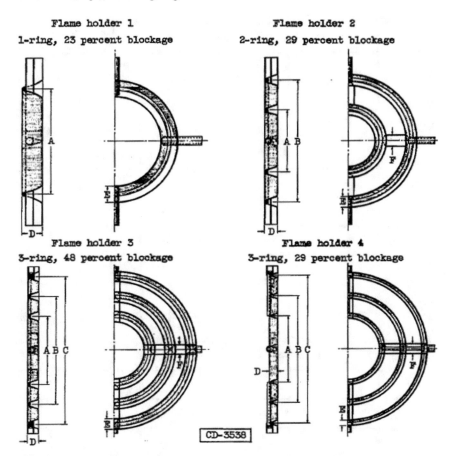

Fig. 6.22 Flameholder annular configurations [18]

Therefore, blockage ratios higher than 35% were disregarded. Efficiency of 93–94% is obtained for the two configurations. Figure 6.26 shows the effect of annular V-gutters width on stability limits. It may be concluded that the operable range improves with the width of the gutters (see Fig. 6.27). Selection of a proper gutter width must be accompanied by proper radial positioning of the gutters with respect to the velocity profile, to ensure satisfactory performance (Fig. 6.28).

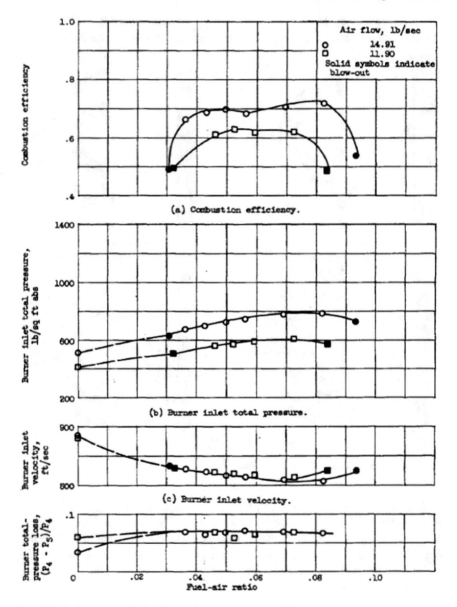

Fig. 6.23 Performance of 1 ring flameholder configuration [18]

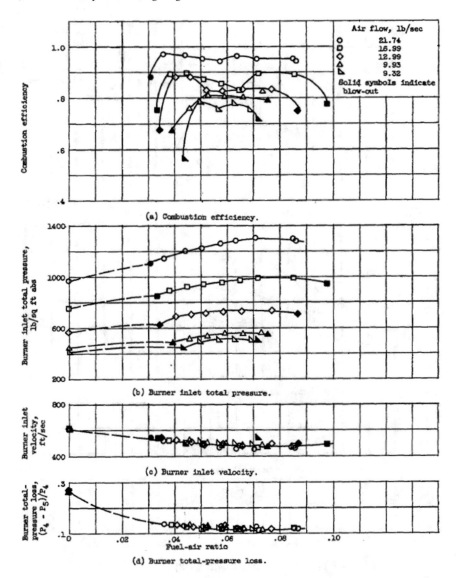

Fig. 6.24 Performance of 3 rings flameholder configuration, $B = 48\%$ [18]

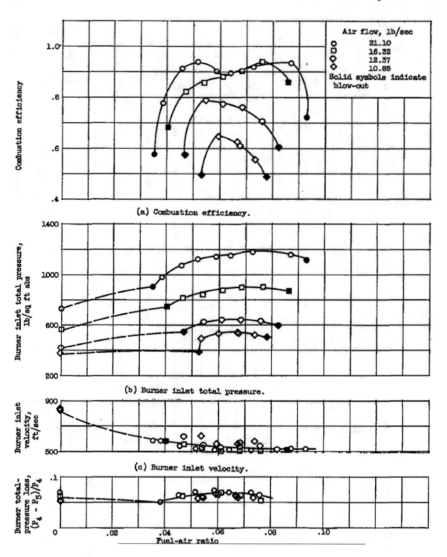

Fig. 6.25 Performance of 3 rings flameholder, $B = 29\%$ [18]

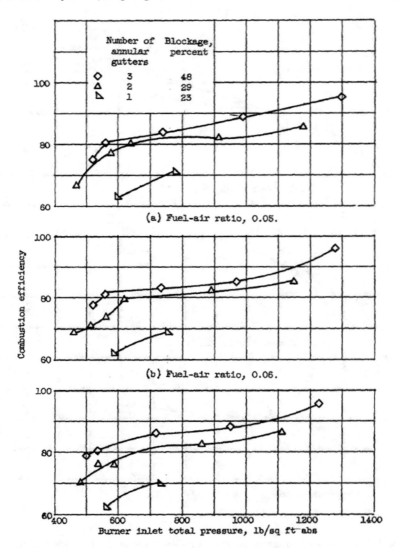

Fig. 6.26 Combustion efficiency of annular V-gutters for different blockages [18]

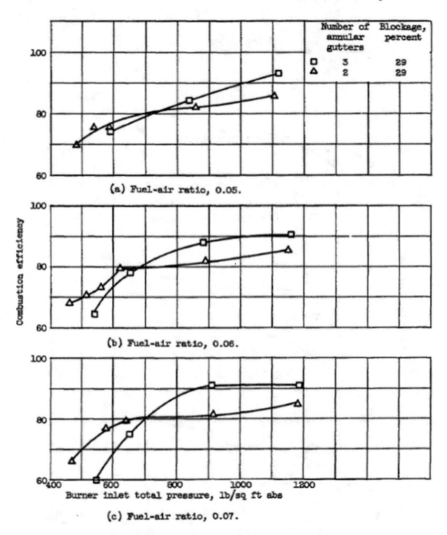

Fig. 6.27 Effect of number of annular V-gutters on combustion efficiency for flameholders of equal blockages 29% [7]

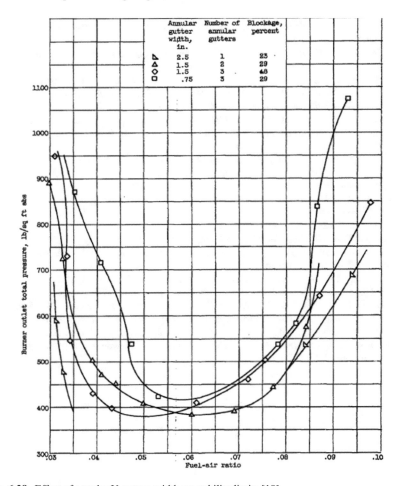

Fig. 6.28   Effect of annular V-gutters width on stability limits [18]

References

1. Effect of three flame holder configurations on subsonic flight performance of rectangular ramjet over range of altitudes, Dugald O.Blach and Wesley E. Messing, NACA RM No. E8I01, 1948
2. Ramjet Engines, M.N. Bondaryuk, F-TS-9740/V 1960
3. E.E. Zukoski, F.E. Marble, Experiments concerning the mechanism of flame blowoff from bluff-bodies, 205–225 (1956)
4. W.H. Sterbentz, Analysis and experimental observation of pressure losses in ram-jet combustion chambers, NACA RM E9H19 (1949)
5. NASA Contractor Report 175064 Spontaneous Ignition Delay Characteristics of Hydrocarbon Fuel/Air Mixtures, Arhur Lefebvre,W. Freeman, and L.Cowell Purdue University West Lafayette, Indiana February 1986
6. J.G. Henzel Jr., L. Bryant, Investigation of effect on number and with of annular flame-holder gutters on afterburner performance, NACA RM E54C30 (1954)

7. S. Nakanishi, W.W. Velie, L. Bryant, An investigation of effects of flame-holder gutter shape on afterburner performance, NACA RM E53J14 (1954)
8. A. Fabrizio, Ramjet performance for different flame holder geometries. Master degree thesis (2014)
9. J.P. Longwell, J.E. Chenevey, W.W. Clark, E.E. Frost, Flame stabilization by baffles in a high velocity gas stream. 3rd Symposium on combustion and flame, and explosion phenomena. The Williams & Wilkins Co., Baltimore, Ma. 40–44 (1949)
10. H. Williams, C. Hottel, A.C. Scurlock, Flame stabilisation and propagation in high velocity gas streams. 3rd Symposium on combustion and flame and explosion phenomena. The Williams & Wilkins Co., Baltimore, Ma. 21–40 (1949)
11. F.H. Wright, Bluff-body flame stabilisation: blockage effects. Comb Flame **26**, 319–337 (1959)
12. A.H. Lefebvre, Gas turbine combustion. McGraw-Hill, series in energy, combustion and environment (1983)
13. D.R. Ballal, A.H. Lefebvre, Weak extinction limits of turbulent flowing mixtures. J. Eng. Power **101**, 343–348 (1979)
14. M. Baxter, A. Lefebvre, Flame stabilization in high-velocity heterogeneous fuel-air mixtures. J. Propul. Power **8**, 1138–1143. https://doi.org/10.2514/3.11454 (1992)
15. P.G. Walburn, Activation energies in a baffle stabilized flame author links open overlay panel. Combust. Flame **12**(6), 550–556 (1968)
16. J. C. Pan, M.D. Vangsness, D.R. Ballal, Aerodynamics of bluff-body stabilized confined turbulent premixed flames. J. Eng. Gas. Turb. Power **114**(4), 783–789 (1992)
17. M.E. Howard, F.A. Wilcox, D.T. Dupree, *Combustion-Chamber Performance With Four Fuels in Bumblebee 18-Inch Ramjet Incorporating Various Rake- Or Gutter-Type Flame Holders* (NACA research memorandum, Lewis Flight Propulsion Laboratory, Cleveland, Ohio, 1948).
18. J.F. Driscoll, C.C. Rasmussen, Correlation and analysis of blowout limits of flames in high-speed airflows. J. Propul. Power, Nov. (2005)

Chapter 7
Injector Design Guidelines

Abstract The primary objective of a liquid fuel injection system is to provide the combustion chamber with the proper amount of fuel and with a spatial pattern that will result in efficient combustion over the entire flight path of the ramjet. In [1], empirical methods for designing fuel injectors were defined. In this work, the approach of Wu [2] to calculate the breakup time of the jet, the correlation of Lefevbre [3] for the average diameter of the droplets of fuel, the study of Turns [4] to calculate the time of vaporization of the droplet and the data processed by Kundu, Colket and Fuller [5–7] for the delay have been selected to provide guidelines to injector design.

7.1 Investigation of Injector Number, Size and Radial Distribution in Flame Holders

The fuel consumption G_g in kg/s is the starting point in determining the number of injectors n, their cross-section, the fuel feed pressure Δp and the density of the fuel:

$$G_g = C_j n_j A_j \sqrt{2 g \rho_f \Delta p_j} \tag{7.1}$$

The flow coefficient C_j depends on the injector geometric characteristics and on the viscosity of the liquid fuel μ_l. With an increase of the viscosity, the swirl of the fuel in a centrifugal injector deteriorates, the thickness of the shroud increases and the flow coefficient grows but still remains <1.

The flow coefficient of direct spray injectors diminishes with the increase of fuel viscosity due to the increase of friction losses. Other conditions being equal, the flow coefficient C_j of centrifugal injectors is smaller than that of direct spray injectors.

The all-important fuel atomization is determined by fuel–air relative velocity, surface tension, and fuel viscosity and density. In fact, defining the Sauter mean diameter (SMD) as the diameter of a sphere that has the same volume/surface area ratio of a particle of interest, that is,

© The Author(s), under exclusive license to Springer Nature Switzerland AG 2021   75
A. Ingenito, *Subsonic Combustion Ramjet Design*,
SpringerBriefs in Applied Sciences and Technology,
https://doi.org/10.1007/978-3-030-66881-5_7

$$SMD = \left(\frac{\sum nD^3}{\sum nD^2}\right) \qquad (7.2)$$

Fuller experimentally proposed to correlate SMD with injection mass flow rate of a single injector, fuel viscosity, pressure drop and air density:

$$SMD = 2.25\pi^{0.25}\sigma^{0.25}\mu_l^{0.25}D_j^{0.5}\rho_l^{-0.25}v_j^{-0.75}\rho_A^{-0.25} \qquad (7.3)$$

Therefore introducing:

$$\dot{m}_j = \rho_l A_j v_j = \rho_l \pi \frac{D_j^2}{4} v_j \qquad (7.4)$$

and the injection pressure drop

$$\Delta p_j = \frac{1}{2}\rho_l v_j^2 \qquad (7.5)$$

within the SMD relation:

$$SMD = \frac{2.25\pi^{0.25}\sigma^{0.25}\mu_l^{0.25}D_j^{0.5}}{\rho_l^{0.25}v_j^{0.75}\rho_A^{0.25}} \qquad (7.6)$$

the strong dependence of SMD on the injector diameter D_j and velocities v_j is evident.

The shape of the jet dispersion depends on the flow velocity, the fineness of the dispersion, and on airstream density and viscosity.

The ratio between the transverse jet area along the combustion chamber and the injector area A_j, i. e., A/A_j is given by Wu [6]:

$$\frac{A}{A_j} = 121q^{0.34}\left(\frac{Z}{d_o}\right)^{0.52} \qquad (7.7)$$

where q is the fuel air momentum ratio, x is the distance from the injector in the direction of air flow and d_0 is the injector diameter (Fig. 7.1).

With an increase of fuel density, a decrease of air density and a decrease of the injection diameter, the dispersion jet widens. As the jet widens, the local fuel concentration in the wake of the injector diminishes, other conditions being equal.

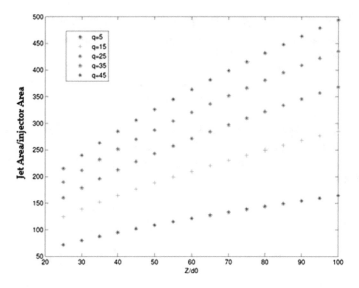

Fig. 7.1 A/A_j versus Z/d_0

7.2 Droplets Evaporation Time

The vaporization of fuel droplets increases with an improvement of the dispersion, with an increase of the volatility of the fuel, an increase of the relative velocity of the drops, the temperature of the air and, especially, the fuel temperature. With an increase of volatility, the vapor concentration increases in the injector wake. Vaporization of fuel occurs where relative velocities are high and therefore along the curvilinear portion of the trajectory after leaving the injector. Therefore, the fuel vapor increases substantially at a very small distance from the injector. With a further increase of distance, vaporization rate increases less than the turbulence mixing rate with air (that is accompanied by a widening of the jet), and the concentration of fuel vapor decreases. The vaporization time of droplets having SMD diameter d_0^2 has been estimated by implemented the Turns correlation:

$$\tau_{ev} = \frac{d_0^2}{K_{ev}} \tag{7.8}$$

where the K_{ev} constant is defined as a function of the thermal conductivity, k_g, the density of the liquid fuel, ρ_l, the specific heat at constant pressure of the gas phase, c_{pg}, and the dimensionless number transfer, B_q:

$$K_{ev} = \frac{8k_g}{\rho_l c_{pg}} \ln\left(B_q + 1\right) \tag{7.9}$$

The dimensionless transfer number B_q is a function of the specific heat of the fuel gas phase, the difference between fuel vapor and liquid temperature, divided by the formation enthalpy of the fuel gas phase:

$$B_q = \frac{c_{pg}(T_\infty - T_l)}{h_{fg}} \tag{7.10}$$

Following the approach of Law and Williams, Turns suggests the following approximations to estimate the specific heat of the gas phase fuel, the difference under of the fuel and the liquid temperature, divided by the formation enthalpy of the fuel gas phase:

$$c_{pg} = c_{pf}(T)$$

$$k_g = 0.4k_f + 0.6k_\infty(T_\infty)$$

where $T = (T_l + T_\infty)/2$.

7.3   Fuel Jet Penetration and Breakup

Fundamental features of jet penetration and disintegration may be identified to take place in three separate regions (Fig. 7.2), that is, (1) the liquid column, (2) the ligaments, and (3) the droplet regions. The formation of the droplet region depends on the ligament region, which in turn depends on the liquid column breakup behavior. Therefore, a thorough understanding of the entire process must begin with complete characterization of the liquid column. The liquid column is the core of jet liquid that forms a continuous stream between the jet exit and the first point of complete fracture. This core may be turbulent or non-turbulent at the nozzle exit and will in either case exhibit distinct surface breakup characteristics. Furthermore, over a wide range of jet operating conditions, liquid properties, and geometrical configurations, the liquid column appearance, length, and steadiness will vary significantly. Estimating fuel jet penetration into an airstream is very important for ramjet-combustor design. Different correlations were proposed by several authors [6], see Table 7.1.

These correlations show that the penetration increases approximately with the square of the fuel momentum. An increase of q increases the penetration and the liquid column and makes the trajectory straighter and thinner.

Table 7.1 Penetration of a fuel jet into an airstream

Chen et al. [7]	$\frac{y}{d} = 9.91(q)^{0.44}\left(1 - \exp\frac{-x/d}{13.1}\right)\left(1 + 1.67\exp\frac{-x/d}{4.77}\right)\left(1 + 1.06\exp\frac{-x/d}{0.86}\right)$
Wu et al. [8]	$\frac{y}{d} = 1.37\sqrt{q(x/d)}$, $\frac{y_b}{d} = 3.44\sqrt{q}$, $\frac{x_b}{d} = 8.06$
Wu et al. [9]	$\frac{y_r}{d} = 4.3\,q^{0.33}\left(\frac{x}{d}\right)^{0.33}$, $\frac{y_m}{d} = 0.51\,q^{0.63}\left(\frac{x}{d}\right)^{0.41}$, $\frac{Z_w}{d} = 7.86\,q^{0.37}\left(\frac{x}{d}\right)^{0.33}$
Inamura et al. [10]	$\frac{y}{d} = (1.18 + 0.24d)\,q^{0.36}\ln\left(1.56 + (1 + 0.48d)\frac{x}{d}\right)$, $\frac{Z}{d} = 1.4\,q^{0.38}\left(\frac{x}{d}\right)^{0.49}$
Becker and Hassa [11]	$\frac{z}{d} =$ $1.48\,q^{0.42}\ln\left(1 + 3.56\frac{x}{d}\right)(q = 1 - 40,\ \mathrm{We_{aero}} = 90 - 2120,\ x/d = 2 - 18)$
Iyogun et al. [12]	$\frac{y}{d} = 1.997\left(q\frac{x}{d}\right)^{0.444}$
Masuda et al. [13]	$\frac{y}{d} = 0.92\,q^{0.50}\left(\frac{x}{d}\right)^{0.33}$
Lakhamraju and Jeng [14]	$\frac{y}{d} = 1.8444\,q^{0.546}\ln\left(1 + 1.324\left(\frac{x}{d}\right)\right)(T_\infty/T_0)^{-0.117}$
Elshamy and Jeng [15]	$\frac{y}{d} = 4.95\left(\frac{x}{d} + 0.5\right)^{0.279} q^{0.424}\,We^{-0.076}\left(\frac{p}{p_0}\right)^{-0.051}$ upper boundary $\frac{y}{d} = 4.26\left(\frac{x}{d} - 0.5\right)^{0.349} q^{0.408}\,We^{-0.30}\left(\frac{p}{p_0}\right)^{0.111}$ lower boundary

Fuller [8] investigated the breaking process of a liquid jet in a subsonic airflow. The parameter of Jet breakup, T_b

$$T_b = \frac{\tau_{ab}}{\tau_{fb}} = 1.52\frac{v_j}{v_A - v_j\cos\theta}\sqrt{\frac{\rho_j}{\rho_A}}\,We_{fd}^{-1/3} \qquad (7.11)$$

is defined as the ratio between the time required for the jet breakup via aerodynamic shear (τ_{ab}) and nonaerodynamic (τ_{fb}) breaking:

Fig. 7.2 Jet penetration and disintegration

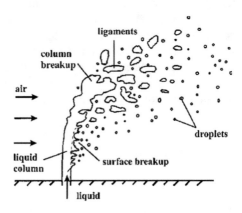

$$\tau_{ab} = 2.58 \frac{D_j}{v_A - v_j \cos\theta} \sqrt{\frac{\rho_j}{\rho_A}} \tag{7.12}$$

$$\tau_{fb} = 1.7 \frac{D_j}{v_j} We_{fd}^{1/3} \tag{7.13}$$

The Weber number is defined as:

$$We_{fd} = \frac{\rho_j d_j v_j^2}{\sigma} \tag{7.14}$$

The transition region according to Fuller [8] is quite limited to $0.9 < T_b < 1.5$; in a first approximation, for $T_b < 1$ the jet breakup is aerodynamic and occurs at a time τ_{ab}; instead for $T_b > 1$ the jet breaks due to the forces in the liquid (surface tension and viscosity) and occurs within a time τ_{fb}.

The distance where the jet breaks is calculated by:

$$\frac{x_b}{d_j} = 9.31 + 2.58 \frac{v_j \cos\theta}{v_A - v_j \cos\theta} \sqrt{\frac{\rho_j}{\rho_A}} \tag{7.15}$$

for $T_b < 1$, i.e., if the breakup is aerodynamics and by

$$\frac{x_b}{d_j} = \frac{9.31}{T_b^2} + 1.7 We_{fd}^{1/3} \cos\theta \tag{7.16}$$

for $T_b > 1$.

These equations show that as the injection angle is decreased, the relative freestream velocity u_{rel} also decreased. When the injection angle is very low, the entire liquid column becomes nearly aligned with the freestream, and the flow conditions approach the case of a parallel jet in a moving stream. In [8], for the case of $M_j = 0.2$, as the injection angle is decreased, the liquid column straightens, the overall penetration decreases within the field, the atomization process is inhibited, and the spray appears to become less uniform. Assuming a higher injection Mach number, i.e., $M_j = 0.4$, with decreasing injection angle, the liquid column straightens, the column penetration decreases, and the atomization process is inhibited. Furthermore, a decrease in the injection angle causes an increase in the axial component of the jet velocity and, therefore, a decrease in the relative velocity between the liquid and the air. This causes the breakup regime parameter T_b to decrease and may cause a transition towards the breakup regime.

7.4 Calculation of Fuel Spatial Distribution with the Distance from the Flameholder

The relative amounts of fuel and air (the spatial distribution of equivalence ratio) are important in determining the flameholder position and the efficiency of combustion. This local fuel/air ratio can be estimated once the penetration and spreading characteristics of the fuel spray are known. Then the proper distance of the fuel injector from the flameholding region is selected to obtain stoichiometric conditions there. The local concentration of the mixture is usually greater than the nominal average. Hojnecki [16] defined a correlation to calculate the local air–liquid fuel ratio, f_l as function of the global air–fuel ratio:

$$f_l = A_{cc}/A_{cjf} \tag{7.17}$$

where A_{cc} is the combustor section, f is the overall fuel air ratio and A_{cjf} is the cross-jet section.

Figure 7.3 shows the local equivalence ratio as function of the distance Z from the injectors. The distance where the equivalence ratio is equal to one is the best position for flameholders.

When the distance between injectors and flameholder is less than that at which the local concentration of fuel ceases to vary, i.e., until the concentration zones become uniform, their location has an effect. In fact, when the flame holder and the injectors are located too close, a narrow jet containing fewer drops and more liquid impinges on the flameholder. These are conditions unsuitable to ignition and combustion.

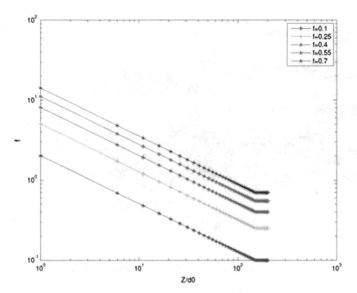

Fig. 7.3 Local equivalence ratio versus Z/d_0

When the flameholder is located further from the injectors, the width of the jet grows, the vapor phase concentration increases, and that of the liquid phase decreases; conditions for ignition and combustion improve and are optimal where the equivalence ratio is stoichiometric. Further increasing of the distance between the flameholder and the injectors slowly widens the vapor plume due to turbulent intermixing; the fuel vaporization rate grows until it finally reaches 100% and there is no additional improvement in combustion efficiency. Varying the longitudinal location of the fuel injectors with respect to flame holders has a little effect on combustion efficiency.

7.5   Experimental Analysis of Injector Radial Distance from the Outer Wall

Several investigators [16, 17] demonstrated that fuel spatial distribution exerts an important influence on combustor performance. At lean overall fuel–air ratios, the fuel distribution was found to have a greater effect than that of flame-holder geometry on combustor performance. In [17], the effect of fuel air distribution on performance of a 40.6 cm ramjet engine, 444.5 cm long (whose combustion chamber and nozzle length is 228.6 cm), and of a 61 cm ramjet were investigated (see Figs. 7.4 and 7.5). The fuel injectors were located 50.8 cm upstream the flameholder. Four fuel tubes entered the engine through the outer wall, each supplying a quadrant injector consisting of four modified commercial spray nozzles. The injectors could be moved radially between the outer wall and the inner diffuser wall (Fig. 7.6).

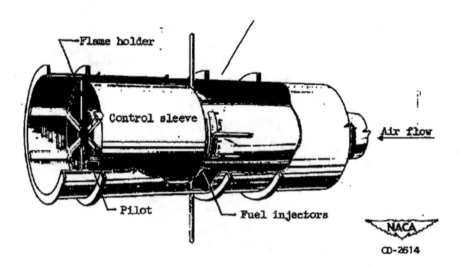

Installation of 16-inch ram-jet engine.

Fig. 7.4 40.6 cm ramjet engine [17]

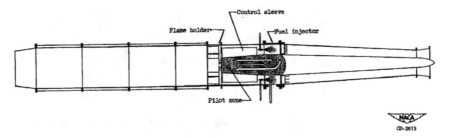

Fig. 7.5 Sketch of the 40.6 ramjet engine showing position of the fuel injectors, flameholder and mixing control sleeve [17]

Fig. 7.6 Flameholder configurations A, B, C [17]

Fuel was JP-3. The flameholders used in that investigation are shown in Fig. 7.7. Configuration A is a grid-type-V-gutter flameholder with a blocked area of 54%, and 3.175 cm across the open end of the V-gutter. Configuration B, is an immersed-surface flame holder with a blocked area of 37%. Configuration C, consists of radial V-gutters with a blocked area of 37%. B and C measured 3.175 cm across the open end. Fuel-air samples were taken at a station immediately upstream of the flame holders. Samples were taken at an overall fuel–air ratio of 0.035 for each flameholder configuration, and for conditions of burning and no burning in the combustor.

The fuel–air survey was made from the outer wall of the ram-jet engine to the inner wall formed by the center body.

A comparison of the combustion-efficiency data obtained with the three flame holders, each with the fuel injector at the same position shows that at a fuel–air ratio greater than 0.03, combustion efficiency was 90–100%. Flame-holder geometry had little effect on the combustion efficiency of the engine despite flame-holder blocked area variations from 37 to 54%. The uniformity in combustion efficiencies obtained with the three flame holders was apparently due to the 588 K inlet air temperatures and higher than atmospheric pressures at which the engine was operated. The drop in air temperature between the fuel injectors and flame holders indicated that fuel vaporization was substantially complete before reaching the flame holders.

In [18], the NACA Lewis laboratory experimentally investigated the combustion efficiency of a 40.64 cm ram-jet engine, operating at conditions simulating a

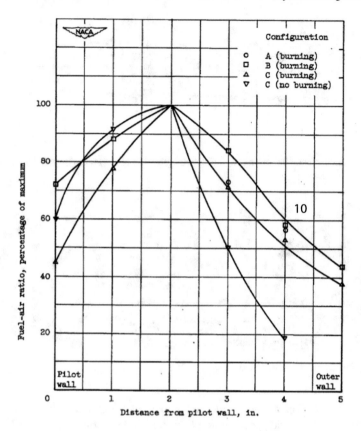

Fig. 7.7 Radial fuel–air distribution upstream of the flame holder [19]

flight Mach number of 2.9. These experiments showed that combustion efficiency was insensitive to fuel preheating and variations in longitudinal location of the fuel injector. Fuel–air surveys indicated that for a fuel temperature of $300 K$ at the injector, 58% of the fuel was vaporized within 15.24 cm of the point of injection.

In [10], it was shown that varying the radial position of the fuel injectors resulted in combustion-efficiency variations of approximately 5%.

As for the lean fuel–air ratio limits, the radial fuel injector position had a small effect on the maximum combustion efficiency but had a pronounced effect on lean blow-out limits. The blowout limits for configuration C, for example, are extended from a fuel–air ratio of 0.0275 with fuel injection near the outer wall to 0.011 with fuel injected near the inner wall.

The radial fuel–air distribution upstream of the flame holder is plotted as a percentage of the maximum fuel–air ratio in Fig. 7.7. Data for each configuration were taken "at an over-all fuel-air ratio of 0.035" and with the fuel "fuel injector at the mid position between the outer and inner wall." For all configurations tested, the maximum fuel–air ratio occurred at the same radial distance from the pilot wall.

However, Fig. 7.7 shows some variation in the fuel–air profiles for the three flame holders investigated.

References

1. A. Mashayek, Experimental and numerical study of liquid jets in crossflow. Master of Applied Science (2006)
2. R.P. Fuller, P.-K. Wu, K.A. Kirkendall, Effects of injection angle on atomization of liquid jets in transverse air flow. AIAA J. **38**(1) January (2000)
3. A.H. Lefebvre, Gas turbine combustion, Hemisphere Publishing Corporation (1983)
4. S.R. Turns, *An Introduction to Combustion*, 2nd edn. (McGraw, Hill international edition, 2000)
5. P. Kundo, VanOverbeke, A practical kinetic mechanism for computing combustion in gas turbineengines, No. 99-2218, 1999, 35rd Joint Propulsion Conference and Exhibit
6. M.B. Colket, L.J. Spadaccini, Scramjet fuels autoignition study. J. Propul. Power **17**(2), 315–323 (2001)
7. T.H. Chen, C.R. Smith, D.G. Schommer, A.S. Nejad, Multi-zone behavior of transverse liquid jet in high-speed flow. 31st Aerospace Sciences Meeting & Exhibit, Jan 1993/Reno, NV (1993)
8. P.-K. Wu, K.A. Kirkendall, R.P. Fuller, Spray Structures of Liquid Fuel Jets Atomized in Subsonic Crossflows. https://doi.org/10.2514/6.1998-714 (1997)
9. P.-K. Wu, K.A. Kirkendall, R.P. Fuller, A.S. Nejad, Spray Structures of Liquid Jets Atomized in Subsonic Crossflows. J. Propul. Power **14**(2), 173–182 (1998)
10. T. Inamura, N. Nagai, T. Watanabe, N. Yatsuyanagi, Disintegration of liquid and slurry jets traversing subsonic airstreams, in *Experimental Heat Transfer, Fluid Mechanics and Thermodynamics*, ed. by M.D. Kelleher et al., pp. 1522–1529 (1993)
11. J. Becker, C. Hassa, Breakup and atomization of a kerosene jet in crossflow at elevated pressure. Atom. Sprays **11**, 49–67 (2002)
12. C.O. Iyogun, M. Birouk, N. Popplewell, H.M. Soliman, Trajectory of water jet exposed to low subsonic cossflow. J. Atom. Sprays (2005)
13. B.J. Masuda, R.L. Hack, V.G. McDonell, G.W. Oskam, D.J. Cramb, Some observations of liquid jet in crossflow, ILASS Americas. 18th Annual Conference on Liquid and Atomization and Spray Systems, Irvine, CA (2005)
14. R.R. Lakhamraju, S.M. Jeng, Liquid jet breakup studies in subsonic air stream at elevated temperatures, ILASS Americas. 18th Annual Conference on Liquid and Atomization and Spray Systems, Irvine, CA (2005)
15. O.M. Elshamy, S.M. Jeng, Study of liquid jet in crossflow at elevated ambient pressures, ILASS Americas. 18th Annual Conference on Liquid and Atomization and Spray Systems, Irvine, CA (2005)
16. J.T. Hojnacki, Ramjet engine fuel injection studies. Air Force report AFAPL-TR-72-76, 1972
17. T.J. Nussdorfer, D.C. Sederstrom, E. Perchonok, Investigation of Combustion in 16-Inch Ram Jet under Simulated Conditions of High Altitude and Eigh Mach Number. NACARME50D04 (1950)
18. T.B. Shillito, S. Naksnishi, Effect of Design Changes and Operating Conditions on Combustion and Operational Performance of a 28-Inch Diameter RamJet Engine. NACA RM E51J24 (1952)
19. J. Cervenka, E.E. Dangle, Effect of fuel-air distribution on performance of a 16-inch ram-jet engine, D. Küchemann, P. Carrière, B. Etkin, Progress in Aeronautical Sciences, Vol. 8, Elsevier, 2 lug 2016

Chapter 8
Combustor Design Guidelines

Abstract A conventional ramjet combustion chamber is simply a hollow cylinder. Combustion chambers may be divided into single stage and two stages according to the organization of their combustion processes. A stabilized combustion chamber includes the following elements: (1) an installation for the introduction and atomization of the fuel; (2) ignition units; (3) turbulence rings; (4) flameholders; (5) mixers. Spark ignition, ignition by incandescent wire, and pyrotechnic ignition are possible means of igniting the fuel mixture. Spray, swirl and pneumatic injectors are used to feed the fuel into the combustion chamber. The design of individual elements depends upon the purpose and upon the dimensions of the combustion chamber.

8.1 Cylindrical Combustion Chamber Sizing

The combustion chambers of ramjet engines may be divided into single-regime, which are intended for operation within a narrow range of fuel mixture, velocities, air pressures and fuel flow rates (cruising conditions); and into a multiple-regime, which are intended for operation over a wide range of velocities and flight altitudes, and consequently, over a wide range of velocities and flow pressures in the combustion chambers, fuel mixtures and fuel flow rates (e.g., interceptor missiles). Combustion chambers may be divided into single stage and two stages according to the organization of their combustion processes. The fuel in single-stage combustion chambers is fed into the entire air flow. These combustion chambers are more suitable for operation with rich mixtures. In two-stage combustion chambers, the air is divided into primary and secondary flows, as in the combustors of gas turbine engines. Fuel is introduced into the primary flow, and combustion occurs at the most suitable composition of the mixtures close to stoichiometric. Later the combustion products are mixed with fresh air, with the goal of lowering their temperature to the required value. Two-stage combustion chambers are intended for operation on lean mixtures.

The apparatus for the injection and atomization of the fuel consists of centrifugal or direct spray injectors. To increase the vaporization and mixing of the fuel with the air, it is advantageous to locate the injectors in the high-velocity areas. The ignition units are usually electric spark plug igniters. To facilitate the ignition system, and to increase the reliability of starting the combustion chambers of ramjet engines, the chambers may be supplied with a pyrotechnic cartridge with an electric squib, similar to those in the combustion chambers of liquid fuel rocket engines of the V-2 type. To facilitate the starting and stabilization of the combustion process, the combustion chambers are usually equipped with a pilot light or preliminary combustion chambers, precombustion chambers and flameholders. The precombustion chamber maintains a powerful, constantly operative jet of flames which ignites the basic mixture. The precombustion chamber is located at the inlet of the main combustion chamber.

The diameter of a ramjet combustion chamber may be calculated from the diffuser Area, assuming a chosen Mach number and applying classic isentropic formulations:

$$\frac{A_c}{A^*} = \frac{1}{M_c} \left[\frac{2}{\gamma + 1} \left(1 + \frac{\gamma - 1}{2} M_c^2 \right) \right]^{\frac{\gamma+1}{2(\gamma-1)}} \tag{8.1}$$

In fact, the combustor geometry is a simple hollow cylinder.
The chamber volume is given by:

$$V_c = A_c \cdot L_c \tag{8.2}$$

For small combustion chambers, the convergent volume is about 1/10 the volume of the cylindrical portion of the chamber.

The combustor length is given by the sum of the length required by the fuel jet breakup, vaporization, mixing times, already calculated in the previous sections.

The length of the combustion chamber has a substantial effect on combustion effi-ciency since the stay (residence) time of the gases in the combustion zone depends on it. With an increase in the length of the combustion chamber, combustion effi-ciency increases and at certain length approaches 100%. Increasing the length of the combustion chamber above, this value does not make sense since with an increase of length, the contact surface of the hot gases with the walls of the combustion chamber and the heat losses through the walls grow. In addition to this, the friction losses of the hot gases on the side of the combustion chamber, and drag, increase. The most suitable length is that maximizing the specific impulse. In [1] a length of 150 mm from the flame holder was not enough for efficient combustion. With an increase of length to 450 mm, the efficiency was 100%. Increasing the length over 450 mm lower the impulse since the efficiency cannot increase further and heat and friction losses increase. A "rule of thumb" is to make the combustion chamber length approximately three times the diffuser entrance diameter.

Combustion efficiency may be expressed by:

$$\eta_c = 1 - e^{-\frac{\tau_{res}}{\tau_{comb}}} \tag{8.3}$$

where τ_{res} is the residence time within combustor and τ_{comb} is given by the sum of the breakup evaporation and ignition delay time:

$$\tau_{res} = \tau_{br} + \tau_{ev} + \tau_{ig} \qquad (8.4)$$

already calculated.

Reference

1. T.B. Shillito, S. Naksnishi, Effect of design changes and operating conditions on combustion and operational performance of a 28-Inch diameter RamJet engine. NACA RM E51J24 (1952)

Chapter 9
Igniter Design Guidelines

Abstract For combustion to occur, it is necessary to provide a minimum initial energy (MIE). This energy has to be calculated to design a proper ignition system.

9.1 Flammability Limits

Flammability limits define the range of fuel/air mixtures where combustion occurs once the ignition system is turned on. The upper flammability limit (UFL) is the limit beyond which the mixture is too rich to burn, the lower limit (LFL) represents the limit beyond which the mixture is too lean. These limits may change depending on the initial temperature of the mixture, in particular, for higher temperatures the range of flammability will be extended, i.e., the upper limit will increase, and the lower limit will decrease.

Reference [1] reported that for most hydrocarbons, the LFL decreases by about 8% for each 100 K increase in the temperature of the mixture. With this rule, mixtures containing an infinitesimal concentration of fuel could sustain the propagation of the flame if the temperature was increased to around 1000 K.

The pressure has generally a negligible effect on LFL, unless too low (i.e., <50 mmHg absolute) to support the propagation of the flame. As for the UFL, this increases as the initial pressure increases [2] (see Fig. 9.1).

9.2 Ignition Energy

For combustion to occur, it is necessary to provide a minimum initial energy (MIE). The minimum energy required may be calculated as the energy required to increase the temperature of a stoichiometric mixture volume of size of the laminar flame thickness, $\delta_l \sim \alpha/v_l \sim l$, to the autoignition temperature T_0:

$$MIE \sim \rho_0 c_{p0} T_o \delta_l^3 \tag{9.1}$$

© The Author(s), under exclusive license to Springer Nature Switzerland AG 2021   91
A. Ingenito, *Subsonic Combustion Ramjet Design*,
SpringerBriefs in Applied Sciences and Technology,
https://doi.org/10.1007/978-3-030-66881-5_9

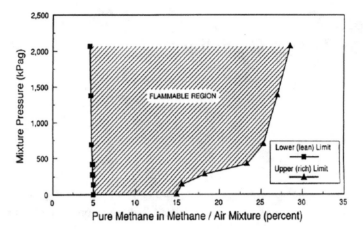

Fig. 9.1 Flammable region versus Pressure and fuel concentration [1]

In general, most mixtures can be ignited by sparks having relatively small energy content (1–100 mJ) but a large power density (greater than $1 MW/cm^3$).

Figure 9.2 shows that the MIE corresponds to the stoichiometric condition, leaner or richer mixtures will require a higher ignition energy.

The goal of the igniter is to turn on the air–fuel mixture. The igniter must generate an electrical arc at high energy (from 4 to 12 J of energy and a few thousands Volt) and must be located in an area of combustion chamber where the fuel and air are already mixed, where the residence time of the mixture is greater than in other parts, and where conditions of temperature and pressure are suitable for combustion, but placed far enough to not be damaged in turn from the high combustion temperatures following ignition.

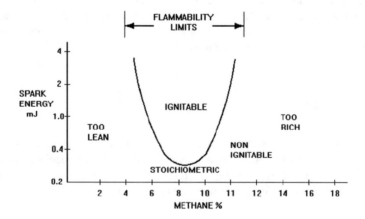

Fig. 9.2 Minimum ignition energy to ignite [2]

If a flow of a fuel mixture, whose composition is within the ignition limits, encounters an ignition source (an electric spark, for example), whose power is insufficient, at a high velocity (about 100 m/sec), then the heat balance proves to be unfavorable, the mixture temperature does not reach T_{ig}, and the mixture does not ignite. To ensure the uninterrupted ignition of the quickly moving flow of a fuel/air mixture, the ignition source (an electric spark, for example) is located behind the flame holder. After ignition, the ignition source may remain active or be disconnected, or carry out the function of a "pilot light." With an increase of the heat energy of the ignition source, the blowout limits at a given velocity and given composition of the mixture are typically expanded.

The Pratt and Whitney J-58 turbojet/ramjet engines powering the Lockheed SR-71 Blackbird spy plane, and its predecessor A-12 OXCART were ignited by means of a tri-ethyl-borane (TEB) igniter. Pyrotechnic ignition of TEB was used to ignite the JP-7 fuel specially formulated to resist autoignition at Mach 3 flight temperatures.

References

1. M.G. Zabetakis, *Flammability Characteristics of Combustible Gases and Vapors, Bulletin 627* (Bureau of nes, Pittsburgh, 1965).
2. C.M. Shu, P.J. Wen, R.H. Chang, Investigations on flammability models and zones for o-xylene under various initial pressure, temperatures and oxygen concentrations. Thermochimica Acta, 371–392 (2002)

Chapter 10
Step by Step RJ Design Methodology

Abstract In this section, an example of this step by step methodology to size a RJ engine for given input parameters and assumptions is reported.

Ramjet input parameters:

- $z = 11000$ m
- $M_0 = 3$
- $\dot{m}_0 = 76.7$ kg/s
- $\Phi = 0.5$

Assumptions:

- $M2 = 0.144$
- $BR = 0.3$
- $\gamma_1 = 1.4$
- $\gamma_2 = 1.3$
- $K = 0.95$

Thermodynamic cycle results

Station 0:

$\quad T_0 = 216.66\ K$

$\quad P_0 = 0.226$ atm

Station 1:

$$T_{t1} = T_0\left(1 + \frac{\gamma - 1}{2}M_0^2\right) = 509.15K$$

$$P_{t1} = P_0\left(1 + \frac{\gamma - 1}{2}M_0^2\right)^{\frac{\gamma}{\gamma-1}} = 9.16\,\text{atm}$$

Station 2:

$\quad \eta_{pd} = \frac{P_{t2}}{P_{t1}} = 1 - 0.075(M_0 - 1) \cdot 1.35 = 0.7975$

© The Author(s), under exclusive license to Springer Nature Switzerland AG 2021
A. Ingenito, *Subsonic Combustion Ramjet Design*,
SpringerBriefs in Applied Sciences and Technology,
https://doi.org/10.1007/978-3-030-66881-5_10

$$T_2 = 507.56$$

$$P_2 = \frac{P_{t2}}{\left(1 + \frac{\gamma-1}{2}M_2^2\right)^{\frac{\gamma}{\gamma-1}}} = 7.31 \text{ atm}$$

$$U_2 = 62.82 \text{ m/s}$$

Station 4:

From CEA code:

$$\frac{T_4}{T_2} = 4.075$$

$$T4 = 2068.74$$
$$P_{t4} = 6.02 \text{ atm}$$
$$M_4 = 0.343$$

Station 9:

$$M_9 = \sqrt{\frac{2}{\gamma-1}\left(\frac{P_{t9}}{P_9}\right)^{\frac{\gamma-1}{\gamma}} - 1} = 2.82$$

$$T_9 = \frac{T_{t9}}{\left(1 + \frac{\gamma-1}{2}M_9^2\right)} = 943.11 \ K$$

$$U_9 = \sqrt{\gamma R T_9} M_9 = 1673.26 \text{ m/s}$$

Performance with $\eta_c = 1$:
$$Ia = 86.068 \text{ s}$$

Pressure losses in combustion chamber due to heat are:
$$\eta_{pc=} \frac{P_{t4}}{P_{t2}} = 0.823$$

Injectors sizing

Assumptions:

- $n_{inj} = 20$
- $q = 20$
- Tsto = 298.15;
- $\sigma = 70.5 \ 10^{-3}$ N/m
- $k_{fuel} = 0.15$
- $\mu_{fuel} = 0.00134$

Results:
$$A_{cc} = \frac{\dot{m}_0}{\rho_2 U_2} = 0.1963 \text{ m}^2$$

$$U_{fuel} = \sqrt{q \frac{\rho_2 U_2^2}{\rho_{fuel}}} = 32.97 \text{ m/s}$$

$$d_{inj} = \sqrt{\frac{4\dot{m}_{fuel}}{\pi n_{inj}\rho_{fuel}U_{fuel}}} = 0.0025 \text{ m}$$

$$\frac{We_f}{1000} = \frac{\rho_{fuel}d_{inj}U_{fuel}^2}{\sigma} = 568.7$$
$$Tb = 0.18$$

$x_b = 0.0493$

$SMD = \dfrac{2.25\pi^{0.25}\mu_{\text{fuel}}^{0.25}d_{\text{inj}}^{0.5}}{\rho_{\text{fuel}}^{0.25}U_{\text{fuel}}^{0.75}P_2^{0.25}} = 9 \times 10^{-5}$ m

$t_{ev} = \dfrac{SMD^2}{k_{ev}} = 0.0257$ s

$L_{\text{phi}} = d_{\text{inj}}\sqrt[0.52]{\dfrac{\left(\frac{A_{\text{jet}}}{A_{\text{inj}}}\right)_{\phi l=1}}{121q^{0.34}}} = 0.063$ m

Flameholder sizing

Input data:

- $BR = 0.3$
- Flameholder number $= 3$
- V gutter angle (alpha_fh) $= 30°$
- Annular configuration
- utsql $= 0.1$

Results:

The flame holder area is:

$A_{fh} = BR_{Acc} = 0.5889$ m^2

$b_{fh} = 0.025$ m

$Lb_{fh} = 0.046$ m

$S_t = 1.61 \cdot S_l \cdot utqsl^{0.742} = 5.81$ m/s

$B_{\text{Lmix}} = \sqrt[3]{0.65 \cdot \dfrac{U_1}{T_2^{0.75}} * \left(\dfrac{BR}{(1-BR)^{0.5}}\right)^{1.5}} = 0.5397$

$L_{\text{flame}} = U_2 \cdot t_{St} \cdot (1 - B_{\text{Lmix}}) = 0.2691$ m

$\phi_{LBO} = \left\{\dfrac{2.25[1+0.4U_2(1+0.1Tu)]}{p_2^{0.25}T_2\left(e^{\frac{T_2}{150}}\right)D_{cc}(1-BR)}\right\}^{0.16} = 0.33582$

For not corrugated chamber, verify screech conditions:

$b_{fh} = 2.5$ cm $< 4\,0.656$ cm $\rightarrow$ condition verified

Ignition sizing

$\tau = 0.0027$ s

Lacc $= 0.0017$ m

$MIE = \rho_2 C_{P2} T_2 L_{\text{acc}}^3 = 42$ mJ

Ramjet design: Cylindrical combustion chamber sizing

The combustion chamber is a hollow cylinder.

$A_{cc} = 0.1963$ m^2

$L_{\text{combreq}} = L_{\text{break}} + L_{\text{evap}} + L_{\text{phi}} + L_{\text{flame}} = 0.42$ m

$L_{comb} = 3 \cdot Dcc = 1.5$ m

$\eta_{\text{comb}} = 1 - exp\left(-\dfrac{L_{\text{comb}}}{L_{\text{combreq}}}\right) = 0.971854$

Iar $= 84.7151$ s

Chapter 11
Material Selection for Ramjet Engines

Abstract Selection of materials for the ramjet engine requires not only considering conventional problems for designing high-temperature materials but also the problem of oxidation and weight. A high degree of reliability combined with a minimum amount of maintenance presents contradictory but necessary design requirements. Combustor accessories such as flameholders require materials having high-temperature strength combined with low oxidation properties. A flameholder must be capable of serving its function without appreciable erosion since burnout could affect the flame-holding stability. Long-range ramjets (cruisers) impose evaluating very critical time-temperature problems. In fact, physical and mechanical properties of materials (modulus of elasticity) vary downward with increasing temperature. The choice of material then is not trivial.

11.1 Operation Temperatures for Materials

Gas turbine technology, which was the most impressive HT development area of the 1950–90 s, progresses at the rate of availability of new materials. Ramjet engines are characterized by lack of rotating parts (compressor and turbine), therefore can operate at combustion temperatures much higher than those of gas turbines and as high as possible. In fact, the ramjet efficiency increases with increasing combustion temperature, thus the ability to operate at high temperature is critical. Typical material temperatures required for different applications are shown in Fig. 11.1.

As for the J-58 afterburning turboramjet installed on the SR-71 Blackbird vehicle, most components were made of Waspalloy (Ni-19Cr-13,5Co-4,3Mo-3Ti-1,5Al [1]), since this is an oxidation-resistant nickel-base alloy capable of withstanding steady 1033 K in air, but burner components were made in Hastelloy-X. Elements with applications similar to those requiring the use of Hastelloy-X (i.e., high temperatures), but which also require greater resistance to buckling and sliding wear, were made of Haynes 25 that is a cobalt-base alloy L-605. Haynes 25 was later replaced with Haynes 188 and Haynes 230, both of which had improved oxidation resistance.

© The Author(s), under exclusive license to Springer Nature Switzerland AG 2021
A. Ingenito, *Subsonic Combustion Ramjet Design*,
SpringerBriefs in Applied Sciences and Technology,
https://doi.org/10.1007/978-3-030-66881-5_11

Fig. 11.1 Operation temperatures for materials in industrial processes [1]

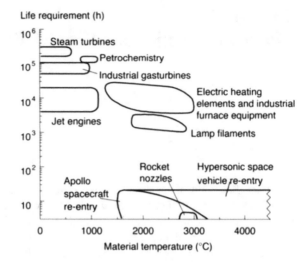

Maximum operating temperatures vary widely, depending on the operative conditions and life. Figures 11.2 and 11.3 show visually the temperature experienced by the JTD-11B-20 (J58) afterburning turbo ramjet installed on the SR-71 Blackbird vehicle.

The next sections show properties and limitations of materials used in high-temperature applications, and materials suitable for ramjet applications are

Fig. 11.2 Picture of J58 engine [2]

TEMPERATURE EXPERIENCED IN THE ENGINE AND NACELLE

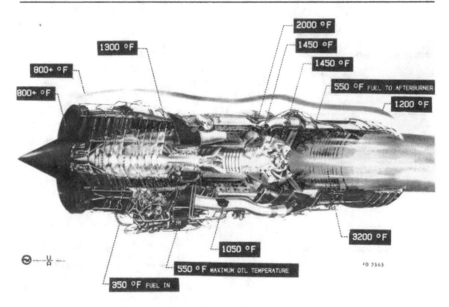

Fig. 11.3 Temperatures experienced by the JTD-11B-20 turboramjet installed on the SR-71 vehicle [2]

suggested. Chronological indication of major industrial material milestones is given in Fig. 11.4 [1].

Among these different materials, stainless steels, superalloys, refractory metals, ceramic, and composite materials are investigated. The stress/temperature capability of the major high-temperature materials systems is compared on a general basis in Fig. 11.5.

Among different materials, Austenitic stainless steel AISI 302 and 321 are often used for high-temperature applications up to 970 K. These steels have low carbon percentage with 18% chromium and 8% nickel and can be machined to manufacture components for combustors [3]. However, these are not particularly suitable for the high temperatures reached in a ramjet. Nickel and cobalt superalloys have good strengths up to 1270 K; they also have good resistance to oxidation and the ability to maintain their characteristics even after long periods of operation at high temperatures. They also have good ductility at low temperatures, useful for the manufacture of the component and an excellent surface stability to chemical attacks and oxidation. However, they are expensive and composed of critical materials such as rare earths. Table 11.1 compares the high-temperature mechanical properties of superalloys and stainless steels. Among most known superalloys [4], there are the nickel-molybdenum steel (Hastelloy X) and chromium-nickel (Inconel) alloys [5]. The superalloy Inconel 718 hardened by precipitation, already used inter alia for turbines, nuclear facilities and cryogenic components and spacecraft, shows good

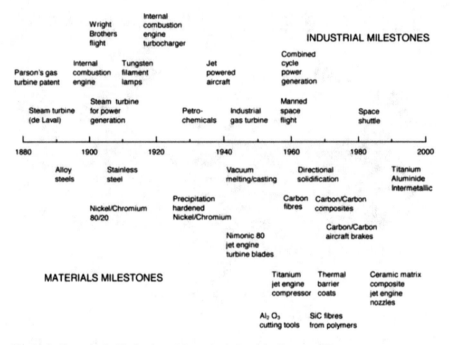

Fig. 11.4 Chronological indication of the major industrial milestones [3]

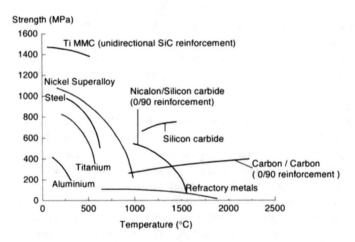

Fig. 11.5 Stress/temperature capability of major classes of high-temperature materials [3]

ductility, creep resistance and fatigue resistance at high temperatures, as well as high mechanical strength, anticorrosion and antioxidation properties. However, studies on the Inconel 718 alloy in high-temperature environments show strength reduction, at a rate which increases with increasing temperature [6]. Moreover, with increasing

Table 11.1 High temperature mechanical properties of super-alloy and stainless steels [4]

Alloys	Elastic modulus (Gpa)	Yield strength $\sigma_{0.2}$ (Mpa)	Creep strength σ_ε^T (Mpa)	Rupture strength σ_t^T (Mpa)
Superalloys				
Carpenter 19-9DL		138 at 815 °C	$\sigma_{1\times10^{-5}}^{732°C}=36$	$\sigma_{1\times10^3}^{816°C}=59$
Incoloy 556™	148 at 800 °C	220 at 760 °C	$\sigma_{1\times10^{-5}}^{760°C}=59$	
Aktiebolag 253 MA	115 at 760 °C	110 at 750 °C	$\sigma_{1\times10^{-5}}^{760°C}=29$	
Haynes R-41	169 at 800 °C	752 at 760°	$\sigma_{1\times10^{-3}}^{732°C}=234$	$\sigma_{1\times10^3}^{816°C}=165$
Inconel 625	160 at 760 °C	421 at 760°	$\sigma_{1\times10^{-3}}^{732°C}=234$	$\sigma_{1\times10^3}^{816°C}=96$
Pyromet 680	144 at 816 °C	241 at 760°	$\sigma_{1\times10^{-3}}^{732°C}=55$	$\sigma_{1\times10^3}^{816°C}=62$
Stainless steels				
AL 446	200 at RT	275* at RT 55* at 760 °C	$\sigma_{1\times10^{-3}}^{760°C}=7.6$	$\sigma_{1\times10^3}^{760°C}=13.5$
Carpenter 443	200 at RT	345 at RT 41 at 760 °C	$\sigma_{1\times10^{-4}}^{704°C}=7.0$	
AL 439 HP™	200 at RT	310 at RT 48 at 760 °C		$\sigma_{1\times10^3}^{816°C}=7.0$

*Minimum as required

relaxation time dislocation density and Vickers micro hardness also decrease and increasingly so with the increase of temperature.

Titanium alloys and some steels can be used to temperatures of about 870–970 K, aluminum alloys cannot be subjected to high loads at temperatures much above about 470 K, but fiber-reinforced aluminum has flexural strengths as high as 300 to 400 MPa at 570 K.

Refractory metals have high strengths at 1470 K, but oxidation resistance is poor. Among refractory metals and their alloys, used for high temperature applications, there are tungsten, tantalum or molybdenum that have an extremely high melting point, but they also have the disadvantage of being expensive and complicated to manufacture.

The characteristics of molybdenum, such as a low coefficient of thermal expansion ($4.8 \, \mu m \, m^{-1} K^{-1}$ at 25 °C) and a high thermal conductivity ($138 \, Wm^{-1} K^{-1}$) [7] make it particularly suitable for environments with large temperature gradients. Its modulus of elasticity is high enough and is not seriously altered by the high temperatures; also has a good creep resistance. To overcome the rapid oxidation of refractory metals, ceramic coatings are the solution of choice.

Tungsten has thermal properties that, as molybdenum, make it suitable for use in combustion chambers. Tungsten is of great interest in many fields, particularly as a structural material in high temperature applications, where it maintains a high ratio strength/weight. However, even for tungsten, usage in these environments depends on coatings applied to the surfaces to prevent oxidation: tungsten oxidation in air begins to be significant at 590 K and increases with temperature up to about 1250 K where degradation increases sharply [8].

Silicide coatings have proved particularly suitable as a protection from the attack by oxygen, but these have a lower limit temperature to 2270 K, which is the temperature where the use of tungsten as the substrate starts to be of interest.

A possible solution to this problem is hafnium–tantalum alloys: they exhibit good ductility when compared with the fragility of many materials for coatings, and also form oxide coatings very adherent and resistant.

For applications to ramjet combustors, high temperature ceramic materials are particularly appropriate, but usually are also brittle, and suitable mainly to bear only compressive loads. The use of ceramic coatings may be advantageous because it allows the use of materials more available compared with metals and may be used in the case where the costs of processing and treatment are excessive; however, ceramic coatings do not provide necessarily thermal insulation.

Studies on high percentage of alumina ceramic materials for ramjet combustion chambers showed that the ceramic surface has flame-holding properties [9], therefore large flow rates can be treated without the risk of blowout, the range of fuel-air ratio is wider, and the efficiency is higher [10].

It is common practice to use composite materials to protect the combustion chambers of the ramjets from high temperatures. Composite materials are becoming increasingly popular, because of their unique properties. These consist of multiple materials of different phases each of which has different chemical properties, divided by an interface. Composite materials have very low densities compared with metals, leading to structures with extremely reduced weight, mechanical properties superior to those of the individual constituents, highly directional and that can stand elevated temperatures. Composite material structures are in fact able to operate in environments with temperatures higher than 1800 K and highly oxidizing [11], provided they are also coated appropriately.

In recent years, composite materials such as the C/C and SiC/SiC, carbon fibers in a carbon matrix, capable of withstanding efforts to traction of 700 MPa and temperatures above 2270 K, have found wide use in aerospace such as parts of the US Shuttle spacecraft and combustion chambers.

Among composite ablative material, the Dow Corning (DC) 93-104 a silicone elastomer material filled with silica and long carbon fibers that provide high performance in environments with large flows and high temperatures, resisting for several minutes at temperatures above 3300 K. Several studies have been made to improve the behavior of this and other composite materials. Results show that one can obtain significant benefits from the addition of diatomaceous earth (or kieselguhr), nanoclay (nanoparticle layers of silicate minerals) and carbon fibers, these providing the best balance between the goals of low heat transfer, high ablation resistance and low density [12].

These properties would lead in principle to decrease the weight of ramjet engines. Also new formulations could potentially reduce the cost of production.

11.2 Ramjet Material Requirements

After reviewing proprieties of the different high temperature materials, the characteristics of materials that can be used for the realization of the combustion chamber of a ramjet are analyzed next [13].

The absolute limiting feature of a possible material is its melting temperature: the high temperatures inside the chamber do not allow the use of traditional materials used for example in jet engines and restrict by far the choice of material used. In practice, the key performance parameter is the yield stress as a function of T. Figure 11.6 is a comparison of the average yield stresses for several structural materials at temperatures up to 1366 K.

Oxidation is the second key factor in designing a combustion chamber, due to the presence of air at high temperature. This problem causes restrictions on the choice of the material that must be able to survive in an oxidizing environment.

Other mechanical and physical properties to be considered to make a thorough examination of all the materials used are:

- The yield stress, which must be high to prevent the material from a plastic-like behavior (irreversible deformation); however, note that stresses in ramjets are moderate
- The density, or, equivalently, the specific weight, a basic property as it affects the weight and thus the design of aircraft.

Fig. 11.6 Yield stresses for structural materials at temperatures up to 1366 K [13]

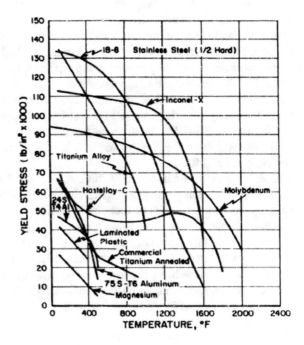

- Low thermal conductivity and low thermal expansion, except wherever different materials are interfaced since thermal gradients and delamination should be avoided.

However, one of the most important factors influencing the choice of materials has little to do with their physical characteristics: the cost.

To design a combustion chamber where walls reach temperatures of about 1300–1400 K, the solutions most used are alloys of chromium, nickel, and cobalt.

The alloys of refractory metals such as niobium, molybdenum, tungsten, and tantalum can, however, work at much higher temperatures since their melting temperature is above 2000 K. The real problem of this type of alloys for ramjet combustors is the high density (niobium, the lightest among the refractory metals mentioned above, has a density of about 8.57 g/cm^3) and the low oxidation resistance. Coatings are therefore required to prevent refractory metals from oxidizing. In this context, zirconia is not suitable as a coating, or as a refractory metal, since oxygen diffuses through the surface oxide layer, and must be stabilized, for instance, with the addition of Yttrium. Therefore, it is the UHTC that should be considered for refractory metal protection. In fact, even though these have higher thermal conductivity than zirconia, thus are less insulating, they may withstand higher temperatures.

At high temperatures, nonmetallic solutions are also interesting to be considered. Ceramic materials, such as Al_2O_3, are characterized by properties at high temperature worthy of note, and often better than those of the super alloys and refractory metals. Their density is much lower than that of a metal alloy, which would make their use in the aeronautical and space far preferable. Their use, however, as already said, is often limited to thermal barriers (TBC-thermal barrier coating) since ceramic materials are very brittle and are not designed to withstand thermal shocks.

Composite materials are also very interesting for high temperature applications. Their properties are closely related to the type of fibers and the matrix used, so it is possible through the use of appropriate materials, obtaining high performance of the composites at elevated temperatures, similar to those of ceramic materials, and characterized by a density lower than those of superalloys and refractory metals. The possibilities are many, but the solutions that seem most promising for ramjet engines (and those which are the focus of research in high temperature) are ceramic matrix composites. or CMC.

CMC materials, for instance SiC/SiC, should also include a coating to prevent carbon oxidation. Ramjet engine material choice is a function of temperature and time since mechanical properties of materials are affected by temperature and load as a function of time. Structural materials must have strength and deformation properties adequate for the intended application.

In ramjet structures, the strength/weight criterion is important. (for a given geometry and loading condition, the material that gives the lightest member is the best choice), see Fig. 11.7. Comparing the specific weight, Hastelloy has the highest when comparing respectively with Inconel, Stainless, Titanium, Aluminum, and Mg.

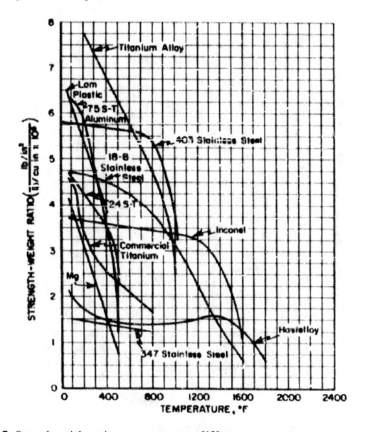

Fig. 11.7 Strength–weight–ratio versus temperature [13]

In high-temperature applications, the strength of a material varies with the temperature, as does the choice of materials based on a strength-weight-temperature criterion.

In the end, the cost versus performance is the most important consideration. Figure 11.8 shows a comparison of strength–cost, ratios versus temperature. It is important to note that this figure does not include fabrication and possible special tooling costs.

The table below can only be used to give a rough estimate of material costs. The numbers are based on a reference about 2002 (Table 11.2).

When producing mechanical components, the material costs generally include production control, forming, machining, finishing, maintenance, corrosion protection and recovery. The price of a metal/metal alloy products results from several factors:

- Alloying grade: certain alloying component can significantly increase the price of the alloy.
- The purity grade: the purer the higher the cost
- Geometry: rolling or forging affects prices per volume or weight

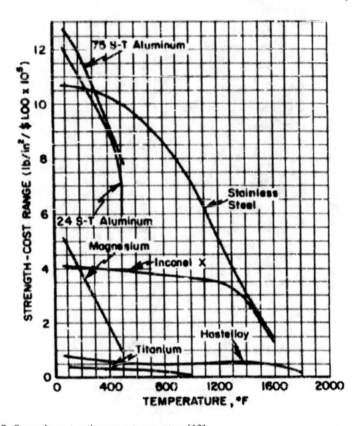

Fig. 11.8 Strength–cost–ratio versus temperature [13]

Table 11.2 Price of raw materials in 2002

Material	Density (kg/m³)	Cost/tonne (£/tonne)	Relative (£/tonne)	Cost/m³ (£/m³)	Relative (£/m³)
Carbon steel	7820	550	1	4301	1.0
Alloy steels	7820	830	1.51	6490.6	1.5
Cast iron	7225	830	1.51	5996.75	1.4
Stainless steel	7780	4450	8.1	34,621	8.0
Aluminum/alloys	2700	2220	4.0	5994	1.4
Copper/alloys	8900	5550	10.1	49,395	11.5
Zinc alloys	7100	2220	4.0	15,762	3.7
Magnesium/alloys	1800	4000	7.3	7200	1.8
Titanium/alloys	4500	17,000	30.9	76,500	17.4
Nickel alloys	8900	18,000	32.7	160,200	36.8

Table 11.3 Price of raw materials in 2010

Material	Cost/tonne (£/tonne)	Relative cost (weight)	Relative cost (volume)
Steel (Billet) LME-Nov-2010	321	1	1
Steel (Hot Rolled Plate)-MEPS-July-2010	505	1.6	1.6
304 Steel (Hot Rolled Plate)-MEPS-July-2010	2,536	7.9	7.9
316 Steel (Hot Rolled Plate)-MEPS-July-2010	3,535	11	11
Tin- LME-Nov-2010	15,458	48	45
Aluminum Alloy—LME-Nov-2010	1,407	4.4	1.5
Aluminum—LME-Nov-2010	1,425	4.4	1.5
Copper—LME -Nov-2010	5,279	16.4	18.7
Zinc—LME -Nov-2010	1,412	4.4	4.0
Nickel—LME -Nov-2010	14,398	44	51
Lead—LME-Nov-2010	1,414	4.4	6.4
Titanium (ingot 6AL-4 V) steel on the net (11$/lb)	15,700	49	28

- Demand: e.g., high defense demand for aerospace industry can result in higher metal prices
- Local economy: e.g., specific metal availability.

For example, in 2010, the cost per ton of HR steel in the form of a 1-diameter, 1 foot long was £3200 against the LME price for the raw material in bulk, that was £321/ton: this illustrates the massive difference in the price of raw materials to the actual price of materials as supplied for initial machining processes in small quantities (Tables 11.3 and 11.4).

Note that 11$/lb is equal to about £15,700/ton. Official prices of raw materials in $/tonne in 2015 are reported in the Table 11.5.

Also in this case, manufactured materials are much more expensive than LME materials. In fact, the molybdenum tube price: was US$ 35,000–150,000/tonne at Nanjing Free-Corrosive Metal Co (Jiangsu, China), that of Inconel 718 N07718 high temperature alloy was US$ 20,000–50,000/tonne and the Haynes 188 round bar/ring was US $ 10,000–40,000/tonne at Shanghai Eraum Alloy Materials Co.

Since the structural material will be subjected to both static and dynamic loads, the strength criterion must be based on the same environment. Hence, the endurance limit of the material will be the appropriate measure of strength. The endurance limit is an indication of the material resistance to fatigue, what for most engineering alloys lies between 1/4 and 1/2 of the ultimate tensile strength of the material, for 10 million or more repetitions of the load.

Table 11.4 Prices in dollars for 25 mm round bar 300 mm long (0.00015 m³)

Metal	Price $	£1000/tonne	£1000/m³	Relative cost (weight)	Relative cost (volume)
Steel HR (A56)	6	3.2	24.7	1	1
Steel CD (12L40)	7.4	3.9	30.5	1.23	1.23
Alloy Steel (4130)	9.99	5.3	41.2	1.66	1.66
St.Stl (304L)	15.34	8.1	63.2	2.6	2.56
St.Stl (316L)	20.61	10.9	84.9	3.4	3.44
Aluminum (2011-T3)	8.58	13.1	35.4	4.2	1.43
Copper (C110)	37.91	17.5	156.2	5.6	6.32
Brass (C360)	25.3	11.7	104.3	3.7	4.22
Bronze	42.09	19..5	173.5	6.2	7.0
Titanium (6AL-4VG5)	107	98	441	31	17.8

Table 11.5 US$/tonne official price (11 June 2015)

(US $/tonne) for 11 June 2015	Official price
Aluminum	1,701.50
Aluminum alloy	1,760.00
NASAAC	1,770.00
Copper	5,906.00
Lead	1,895.50
Nickel	13,465.00
Tin	15,100.00
Zinc	2,124.00
Cobalt	30,250.00
Molybdenum	16,300.00
Steel billet	150.00

For ferrous metals, the endurance limit is considered to approach a limit termed the fatigue strength, which is independent of the number of stress repetitions. This limit disappears at elevated temperatures where, for all materials, the endurance limit is a function of the number of applications of the load. The measure of a material's resistance to static elastic deformation is the modulus of elasticity. This relationship, which approximates the ratio of stress to strain in the elastic range as a straight line, is also a function of temperature and decreases in value with an increase in temperature as shown in Fig. 11.9.

For dynamic deformations, in addition to the modulus of elasticity, the density of the material must be considered, since accelerated masses produce inertia forces.

An ideal material for dynamic applications should be one as light as magnesium and as stiff as tungsten, but the selection of a material for a particular use in usually based on static rather than dynamic deformations.

Fig. 11.9 Modulus of
elasticity versus temperature
[13]

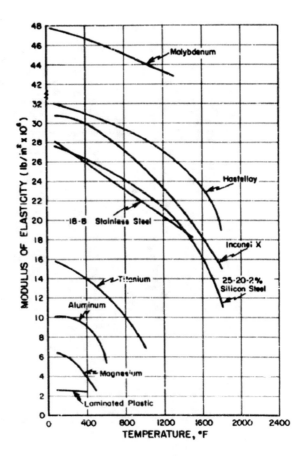

For ramjet engines, these properties increase in importance with longer missile flight time. The problems of thermal expansion and thermal shock are especially important in structures involving several different materials.

Consideration of corrosion problems in military naval applications requires careful selection of galvanically adjacent materials unless extraordinary protection precautions are taken. It is preferable to use materials as close to each other as possible in the galvanic series, and in no case separated by more than one grouping, see Table 11.6.

Another mechanical property of considerable interest is ductility. Since the nature of missile structures leads to combined stresses and strain concentrations, materials with low notch sensitivities are desirable. A usable, but not complete measure of this property is ductility, which shows up in the mechanical properties most commonly as percent elongation in a specified gage length, and less commonly, but more accurately as percent reduction of test specimen area.

Table 11.6 Galvanic series of different materials [13]

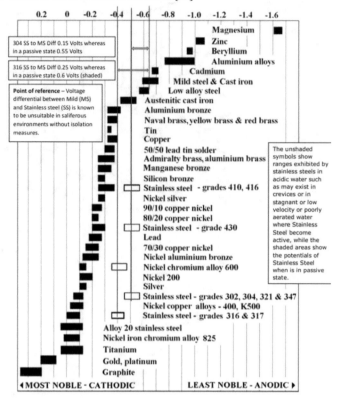

These properties are also some measure of a material's "toughness," which indicate its ability to absorb energy before fracture and is therefore also an important measure of workability.

Any material has certain advantages and disadvantages in a given situation; however, the material selected will usually be a compromise based on the following:

1. It is satisfactory for statistical loads,
2. it is satisfactory for dynamical loads,
3. It fulfills special requirements (such as hot oxidation resistance, damping capacity, etc.)
4. it is practical to fabricate on a production basis,
5. it is available (and is not a critical material in time of war).
6. it is affordable.

Besides cost, a means of rating strategic and critical materials is frequently of value in performing both preliminary and production design. In determining which one of several suitable materials should be selected for a specific application, consideration should be given to the current and possible future scarcity of the material.

There is a method of rating an element numerically according to its relative scarcity by applying material factors ("MF") to the element. Boron, which is considered abundant and noncritical, is given an "MF" of 1, while niobium (in the US: columbium), considered the most scarce and critical, is given an "MF" of 2200.

References

1. E. Pope, *Rules of Thumb for Mechanical Engineers, Accurate Solutions to Everyday Mechanical Engineering Problems* (Gulf Professional Publishing, 1997)
2. Design of High Temperature Metallic Components, ed. Hurst, R.C., Elsevier applied Science Publishers, London, New York (1984)
3. M. Cemal Kushan, S. Cevik Uzgur, Y. Uzunonat, Fehmi, ALLVAC 718 Plus™ Superalloy for Aircraft Engine Applications, Recent Advances in Aircraft Technology (2014)
4. E. Loria (ed.), *Superalloy 718, 625 and various derivatives* (The Minerals, Metals and Materials Society, 1994), pp. 711–720
5. S. Azadian, L.Y. Wei, R. Warren, Delta phase precipitation in Inconel 718 Mater Charact **53**, 7–16 (2004)
6. Climax Mo.Co.: Molybdenum metal (1960)
7. Tungsten: Properties, Chemistry, Technology of the Elements, Alloys, and chemical compound, Erik Lassner, Wolf-Dieter Schubert, Kluwer Academic/Plenum Publishers, 1999
8. L. Zhang, D. Jiang, High Temperature Ceramic Matrix Composites 8, John Wiley & Sons, 19 mag 2014
9. A guide to the control of high temperature corrosion and protection of gas turbine materials, Duret-Thual, C., Morbioli, R., Steinmetz, P., EUR 10682 EN (1986)
10. Composite Materials Handbook, Ceramic Matrix Composites, Volume 5, MIL-HDBK-17-5, 17 June 2002
11. M.R. Begley, N.G.H. Wadley, Delamination of ceramic coatings with embedded metal layers. J. Am. Ceramic Soc. **94**, 96–103 (2011)
12. V. Levitin, *High Temperature Strain of Metals and Alloys: Physical Fundamentals*, John Wiley & Sons, 12 mag 2006
13. C.W. Besserer, Materials for ramjet engines and components, AD036272 Report, APL, 1967

Chapter 12
Conclusions

In this work, a methodology for the ramjet sizing has been proposed. Mission requirements (Altitude, Range, payloads) have been correlated with the engine thrust, and the vehicle mass budget and sizing. Different fuels used for ramjet applications have been investigated and compared in terms of thermochemical proprieties, performance, and reactivity. The effect of flameholder geometry on combustor performance has been analyzed. This analysis has shown that higher blockage ratio allows a wide range of stability limits in terms of fuel–air ratio and inlet air speeds. The geometry has a minor effect on combustor efficiency, in fact, increasing the distance and the size of each gutter does not increase significantly the combustion efficiency. Instead, a critical parameter is the width of the gutter: the operable range improves as the width of the gutter increases. The best position for flameholders is where the fuel–air local concentration is near stoichiometric. Further increasing the injectors to flameholders distance has no effect on combustion efficiency. The radial fuel position variation has an impact of maximum 5% on combustion performance. Empirical laws for the fuel injection system fuel jet penetration as function of the fuel to air momentum ratio, temperature, density and injector diameter have been proposed, as well as empirical laws to calculate the jet breakup, fuel evaporation time and distance. The transverse jet area as function of the distance from injectors has been correlated to the local fuel concentration. Once the overall length to complete combustion is calculated, a preliminary sizing for a cylindrical combustor chamber has been proposed. The energy source required for the initial ignition has been estimated to be of order of 1–100 mJ and about 1 megawatt/cm^3 for air/hydrocarbon mixtures; however, in actual practice ignition depends on the mixture temperature, the total energy supplied, the rate at which energy is supplied, or time period over which it is delivered, and the volume over in which energy is delivered. Igniters are generally located behind the flame holder, where fuel conditions are within limits of flammability. Finally, materials to be used for different components have been listed, their properties analyzed in the context of designing ramjet engines, and potential candidates have been proposed.

A. Ingenito, *Subsonic Combustion Ramjet Design*,
SpringerBriefs in Applied Sciences and Technology,
https://doi.org/10.1007/978-3-030-66881-5_12

Printed in the United States
By Bookmasters